PLACING

OUTER

SPACE

EXPERIMENTAL FUTURES

TECHNOLOGICAL LIVES, SCIENTIFIC
ARTS, ANTHROPOLOGICAL VOICES

a series edited by MICHAEL M. J. FISCHER
and JOSEPH DUMIT

PLACING

OUTER

SPACE

AN EARTHLY ETHNOGRAPHY
OF OTHER WORLDS

LISA MESSERI

Duke University Press   Durham and London   2016

Book and cover design by Natalie F. Smith

Typeset in Quadraat Pro by Tseng Information Systems, Inc.

Printed and bound by CPI Group (UK) Ltd, Croydon, CR0 4YY

Library of Congress Cataloging-in-Publication Data

Names: Messeri, Lisa, [date] author.

Title: Placing outer space : an Earthly ethnography of
other worlds / Lisa Messeri.

Other titles: Experimental futures.

Description: Durham : Duke University Press, 2016.

Series: Experimental futures : technological lives, scientific
arts, anthropological voices | Includes bibliographical
references and index.

Identifiers: LCCN 2016011263|

ISBN 9780822361879 (hardcover : alk. paper) |

ISBN 9780822362036 (pbk. : alk. paper) |

ISBN 9780822373919 (e-book)

Subjects: LCSH: Ethnology. | Planets. | Extrasolar planets.

Classification: LCC GN320 .M574 2016 | DDC 523.2/4—dc23

LC record available at http://lccn.loc.gov/2016011263

Cover art: (top) Hubble Space Telescope image. Credit: NASA,
ESA and the HST Frontier Fields team (STScI); acknowledgement:
Judy Schmidt. (bottom) Opportunity's view of Wdowiak Ridge
(stereo), Sept. 17, 2014. Credit: NASA/JPL-Caltech/Cornell Univ./
Arizona State Univ.

CONTENTS

A C K N O W L E D G M E N T S

I audaciously claim sole authorship of this book, but truly it is a work influenced, aided, and dependent on the graciousness of many others. I am indebted to the many researchers who opened up their labs and lives to me. The MIT exoplanet community kindly allowed me and my notebook to attend their meetings. Sara Seager was not only an informant but a mentor as well. Josh Winn, Leslie Rogers, Sukrit Ranjan, and Nikku Madhusudan shared their research and their wonder with me. It was serendipitous that Debra Fischer was on sabbatical in Cambridge when I was looking for a guide to take me to an observatory. Not only did she invite me, a stranger, on a research trip but she offered friendship and support throughout my research. Carol Stoker similarly took the risk of inviting me to live in close quarters with her for two weeks at the Mars Desert Research Station. Michael Broxton found a home for me at NASA Ames Research Center, and I thank Terry Fong for giving me a desk in the Pirate Lab and Zack Moratto, Ted Scharff, Mike Lundy, Ross Beyer, and Ara Nefian for their camaraderie. Thanks also to the numerous scientists at NASA and those who, when visiting MIT, patiently and engagingly answered my questions in interviews and informal conversations. Thanks also to Glenn Bugos and April Gage, who run the history office at NASA Ames.

The institutions that provided me with intellectual homes during research, writing, and revision shaped how I interpreted the work of these scientists. This book's life began in the classrooms and faculty offices of MIT's Program in History, Anthropology, and Science, Technology, and Society (HASTS). Mentors in this phenomenal program encouraged me to experiment with disciplinary and theoretical approaches, finding means and language to address the puzzles I encountered among space scientists. Stefan Helmreich changed how I saw the world. Not only does he continue to inspire me as a scholar but he is also an exemplar of how to be impossibly generous with ideas and time for students. David Kaiser and David Mindell expertly guided my maneuvering between technical and social fields. Graham Jones arrived as I was leaving MIT, but he gamely stepped aboard my dissertation committee and diligently read and commented on my work. I would also like to thank other teachers and mentors I encountered in Cambridge, including Heather Paxson, Chris Walley, Mike Fischer, Anne McCants, Susan Silbey, Vincent Lépinay, Natasha Schüll, Roe Smith, Roz Williams, Leo Marx, Sheila Jasanoff, Peter Galison, and Peter Gordon. Faculty mentors are all well and good, but the graduate student community of HASTS made the experience joyful. My thanks to Orkideh Behrouzan, Etienne Benson, Laurel Braitman, Nick Buchanan, Marie Burks, Candis Callison, Peter Doshi, Kieran Downes, Amah Edoh, Xaq Frolich, Chihyung Jeon, Shreeharsh Kelkar, Shekhar Krishnan, Tom Özden-Schilling, Canay Özden-Schilling, Anne Pollock, Caterina Scaramelli, Ryan Shapiro, Peter Shulman, Ellan Spero, Michaela Thompson, Ben Wilson, and Sara Wylie as well as non-HASTSies Abby Spinak and Katherine Dykes. As we have gone on to new institutions, Emily Wanderer, David Singerman, Teasel Muir-Harmony, Alma Steingart, Sophia Roosth, and Michael Rossi continue to be great friends to think with. Rebecca Woods is responsible for helping me figure out any clever title and many clever ideas in this book and to her I owe a great thanks for being a constant source of amusement and support.

As a postdoctoral fellow teaching with the Integrated Studies Program at the University of Pennsylvania, Peter Struck gave me a unique home from which to think about the pedagogical and scholarly power of interdisciplinarity. The faculty and students of the History and Sociology of Science Department warmly welcomed me into their lively conversations, and scholars across the university influenced how I approached transforming the dissertation into a book. I thank John Tresch, Greg Urban, Ruth Schwartz Cowan,

Robert Kohler, Matt Hersch, Lisa Ruth Rand, Joanna Radin, Mary Mitchell, Britt Dahlberg, and Beth Hallowell. With two other postdocs, Kate Mason and Jessica Mozersky, I was part of a makeshift cohort, and we helped each other make sense of where we were and where we were going.

For me this meant continuing my southerly migration and finding a home at the University of Virginia in the program of Science, Technology, and Society in the Department of Engineering and Society. Here Bernie Carlson, Mike Gorman, Tolu Odumosu, Rider Foley, Caitlin Wylie, Kay Neeley, Deborah Johnson, and colleagues in the Department of Anthropology, including Susie McKinnon, Kath Weston, Jim Igoe, Ira Bashkow, and China Scherz supported the final shaping of the project. Walks with Deborah, Susie, and Geeta Patel were particularly clarifying. I also thank Sarah Milov, Rachel Wahl, Blaire Cholewa, and Melissa Levy for needed distractions in this final stage.

I was fortunate to become part of a virtual institution of like-minded scholars who are passionate about understanding the sociality of outer space. The impact of writing, conversing, and collaborating with this group is evident on every page of this book. I thank Janet Vertesi, David Valentine, Debbora Battaglia, Zara Mirmalek, Teasel Muir-Harmony, and Lisa Ruth Rand. Valerie Olson has been an especially important interlocutor, and I thank her for the many phone calls and conference coffees.

Fieldwork for this book was carried out with support from MIT's Kelly-Douglas Fund and a doctoral dissertation grant from the National Science Foundation (SES-0956692). While finishing my dissertation, I was partially supported by the Andrew W. Mellon Foundation as a fellow for a John E. Sawyer Seminar on the Comparative Study of Cultures hosted by MIT's Department of Anthropology.

Ken Wissoker wisely advised me throughout this processes. I owe much thanks to him, Elizabeth Ault, Sara Leone, and Duke University Press for supporting this book. The three anonymous reviewers encouraged me to make some significant changes and sagely advised ways to sharpen and strengthen this text. As this process came to a close, I was delighted to be included in Mike Fischer's and Joe Dumit's Experimental Futures series, and my thanks go to them as well.

The discussion of astronaut geology training from chapter 1 appears in an expanded form in *Astropolitics: The International Journal of Space Politics and Policy*.

Many friends and family members have offered enticing, fun reasons to take a break from work. For all of the laughter, I thank Rachael Lapidis, Melissa Read, Teresa Kim, Anna Dietrich, Jeff Roberts, Abby Berry, Laura Gibson, Emily Brennan, Susan Choi, and Jeff Ebert. Armfuls of love go out to my brother, Jason, and family in Maplewood, Verona, and Avalon. Finally, for my parents, Ellen Musikant and Peter Messeri, thank you, simply, for everything.

FROM   OUTER   SPACE

TO   OUTER   PLACE

Concentrating on space, one encounters place.
—Peter Redfield, *Space in the Tropics* (2000)

Two young boys point up at the night sky, silhouetted against a lake reflecting the oranges, blues, and purples of a sky at sunset. Even though we cannot see their faces, their body language speaks an animated excitement. A viewer of this scene, caught on camera by the boys' mother, might wonder what the kids are pointing to and why they are so excited. Sara Seager, the mother, an MIT professor of planetary science and MacArthur fellow, shared this picture with an audience at a conference on exoplanet astronomy, the study of planets orbiting stars other than the Sun. I sat among the astronomers, listening and watching as Seager built off of the energy and aspiration of the picture as she asked the audience to imagine that this picture was taken in the future. What might the boys be pointing to with such excitement? For the assembled audience, the answer was obvious: they will be pointing to a star known to have a planet just like Earth. Most known exoplanets are exotic and strange, but Seager and her colleagues hope that the future of their young field lies in the study of familiar, Earth-like planets. In searching for connections between Earth and other planets, today's planetary scientists refigure the night sky as teeming with *worlds*.

A few years after this conference, I listened to a radio interview with an exoplanet astronomer who had attended Seager's talk. She was trying to convey the significance of her field, asking the host: "When you look up in the sky . . . what do you feel? . . . There's a profound sense of loneliness, I think, that the universe is so big and I'm so small." Exoplanet astronomers claim, however, that their field is changing this structure of feeling. This scientist went on: "Imagine in the near-term future, you know, your grandchild or your great grandchild and his mother can point to a star and say, 'That star, that star right there has a planet just like Earth and it harbors life.' That's a different perspective" (Batalha 2013). This is a perspective that positions Earth not as a singular blue marble floating in a sea of darkness but as one planet among many on which humans might be capable of living.

Though imagining human (and other) life beyond Earth has long been a feature of speculative fiction, scientists are only recently offering definitive proof that such potentially habitable worlds exist and, moreover, might be common. Planetary scientists find meaning in these new discoveries by imagining and talking about planets as *places*. Places on Earth can be cities or villages, landmarks or landscapes. They have a specific character that might change over time or be differently perceived from person to person. But, importantly, one can *be* (or can imagine *being*) in a place. Place suggests an intimacy that can scale down the cosmos to the level of human experience. To claim that the infinite field of stars is not an invitation to loneliness but a prompt for feeling the pull of cosmic companionship requires a powerful and pervasive understanding of these planets as *worlds* and *places* that relate to our own—that invite being. This book is about how this understanding is developing in exoplanet astronomy and studies of our own solar system. It is about how outer space is being made rich with place.

The photo at the conference and the verbal description of a similar image during the radio show might be dismissed as popularization or performance. But these pronouncements go to the core of new ways of thinking about and doing planetary science. Place-making has become central to daily work in the field. Transforming numerical data from telescopes and satellites into full-blown worlds—into places a scientist can imagine visiting—structures the research of today's Mars scientists and exoplanet astronomers alike. The scientists whose works and lives I document in this book metamorphose the dark expanse of the night sky into a zone of fresh meaning and insight. In the face of this grand canvas and millennia of mythologizing the secrets held by the universe, place becomes a tool by which scientists can grapple with their objects of research on a human scale. Place is more than a given category; it is a way of knowing and of making sense. In connecting the mundane and the extraordinary, extraterrestrial place-making grounds knowledge of other planets in familiar contexts. Scientific practices of place-making turn the infinite geography of the cosmos into a theater dotted with potentially meaningful places that are stages for imaginations and aspirations. At the same time that the field of exoplanet astronomy is expanding, Mars science is also enjoying renewed interest. Successful rover missions and satellites returning beautiful, detailed pictures of the surface invite an intimacy with our neighboring planet. In this book, I discuss both Mars and exoplanet research to explore a full range of place-making practices that stretch from a planet fully mapped and photographed to ones with surfaces that are still mysterious and invisible. Exoplanet astronomers and Mars scientists use a variety of methodologies and technical languages and ask distinct research questions, but their shared pursuit of planetary place marks their fields as more similar than different. In these pages, I refer to the work performed by the scientists I encountered in my ethnographic work, be they planetary geologists, astrophysicists, or computer scientists, as part of the larger project of planetary science.[1] Such a frame of reference permits me to emphasize the shared concerns, desires, and beliefs that has made place a sought-after objective in this broader community.

I conducted this ethnography without ever leaving Earth, but my field of study extends well throughout the galactic neighborhood (see fig. Intro.1). It is largely based on fifteen months of participant observation, conducted in 2009 and 2010. Though I entered the field as an anthropologist, I brought with me my undergraduate training from MIT, where I had earned a bachelor of science degree in aeronautical and astronautical engineering. Due to this degree in "rocket science," I was already part of and thus had easy access to the network that my ethnographic work sought to examine. My informants trusted me to participate in the research I was simultaneously observing. I worked with university astronomers in their offices and at a Chilean observatory. I joined Mars scientists in their NASA laboratories and as they traveled into the Martian-looking landscape of the American West. Through interviews, involvement in research projects, conference attendance, chats over beers and pisco sours, and email exchanges, I traced

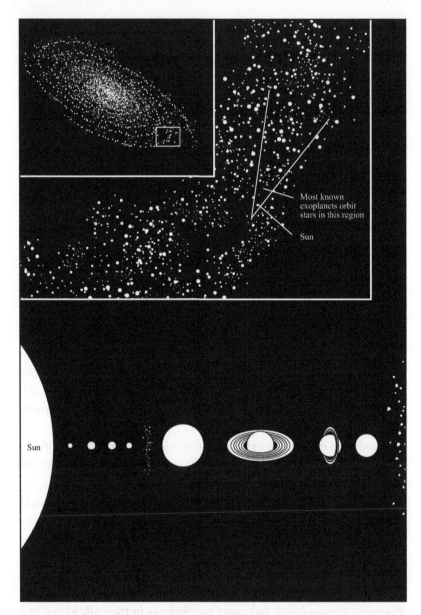

Intro.1 A readers' guide to the universe: map depicting the Milky Way galaxy, the Sun's position in the Orion arm of the galaxy, and the primary bodies of our solar system. This book spends significant time on Earth (third planet from the Sun) and slightly smaller Mars (fourth planet). Note in the middle inset that the 1,000-plus planets orbiting other stars thus far detected are in a relatively small portion of the galaxy. This is the region of exploration afforded by our current technologies, namely the Kepler Space Telescope (see chapter 4). Image credit: Michael Rossi.

how planets are changeable objects, made more meaningful and relatable with each new data set, scientific paper, and conversation. The work of creating planetary place is similar to the anthropologist's own desire to make the strange and alien familiar. The planetary scientists I write about in this book are literally world-builders. They are invested in questions of what it is like to be on other worlds. The ease with which they can imagine such being ultimately makes this book also a story about the changing ways of being in this world.

To provide background for understanding the practice of today's planetary scientists, I will elaborate on the development of this multidisciplinary field and offer a framework for thinking about planets as both scientific objects and worlds. To think at the scale of worlds requires the employment of what I refer to throughout the book as "the planetary imagination." After laying the foundation for this concept, I introduce how place can aid in bridging the planetary scale of such an imagination with our smaller-scale experience of being in the world. I conclude this introduction with an overview of the tour through the galaxy that this book will take and the terrestrial locations that will facilitate this journey.

"The New Interdisciplinary Science of the Solar System"

The field of planetary science originates in the early years of the American space age, when setting foot on other planets seemed a likely future. In 1962 Elsevier began publishing a new journal, Icarus, to document "the new interdisciplinary science of the solar system—which is emerging to claim its own identity at the cross-roads of the allied disciplines of astronomy, geology, geophysics, meteorology, geochemistry, plasma physics, and biology" (Kopal and Wilson 1962, i). The journal established the "planet" as a central object of study, trumping disciplinary divisions in favor of gathering together a diversity of views that could make sense of the astronomical category to which Earth, in the post-Copernican world, belonged. The preface to the first volume of Icarus described the mythological inspiration for the journal's title. Readers were not to be reminded of Icarus's fate but inspired by his dream.[2] Quoting revered astronomer Arthur Eddington, "Icarus will strain his theories to the breaking-point till the weak joints gape." And though he might not have reached his destination, "we may at least hope to learn from his journey some hints to build a better machine" (iii). *Icarus* established the beginning of an ambitious scientific journey, inviting contributors to take risks so that the field could succeed. Scientists did indeed rally behind this new discipline, and by the end of the decade members of the American Astronomical Society had organized within it a subgroup called the Division for Planetary Science.[3]

Shortly after *Icarus*'s first issues circulated among the scientific community, Carl Sagan, then a young assistant professor of astronomy at Harvard, was making a name for himself as both a dynamic teacher and an engaging public speaker. He gave public lectures at the Harvard College Observatory titled "Planets Are Places." This was a new way of viewing the night sky, one in which planets were not points of light identified alongside constellations but, as one of his first graduate students put it years later, "worlds in their own right" (quoted in Davidson 1999, 170). To imagine planets as worlds became easier as the decade progressed and probes were launched to Venus and Mars and Project Apollo made the Moon into a destination. Sagan remained an advocate for imagining planets as places and did so while serving as an associate editor of *Icarus* from its conception and continuing on as an editor-in-chief beginning in 1968.

This first decade of *Icarus* contained more articles about the Moon than any other solar system object, as astronauts and cosmonauts raced to be the first on its surface. However, from the mid-1970s until the present, Mars came to dominate the journal's pages, thanks initially to the successful Mariner and Viking missions of the 1960s and 1970s. Even in the lull between Viking and the next successful Mars surface mission, Mars Pathfinder (launched in 1996), there were enough data on and interest in Mars to keep discussions about it alive in the pages of *Icarus*. In 2009, after more than a decade of successful robotic missions, a record ninety-one articles about Mars were published in *Icarus* (which has remained the premier planetary science journal).

When the field of planetary science was established, scientists assumed that it would be devoted to the solar system. Though there had always been talk and speculation about planets orbiting other stars, there was no robust way to detect such objects. In the 1970s a half dozen articles in *Icarus* addressed the feasibility of detecting exoplanets (or "extrasolar planets" as they were initially called) and how Earth-like planets might be found. It was not until the 1990s, however, that astronomers were finally able to prove the existence of exoplanets. Scientists found the first few around pulsars, violent stars that bathe their companion planets in X-rays. In 1995 Swiss astronomers announced the discovery of a planet around a star very similar to the Sun. After decades of wondering if exoplanets could be detected, the affirmative answer asked, in turn, "How many are there?" and "What are they like?" In October 2013 astronomers confirmed the detection of the one-thousandth exoplanet, and the number grows steadily as astronomers continue to scour our small corner of the Milky Way galaxy.

With this abundance of exoplanets, so has grown their presence in *Icarus*. Though nowhere near as prevalent as those about Mars, there have been consistently more than a half dozen articles about exoplanets per year since 2000. In astrophysics journals, however, articles about exoplanets steadily increased after the 1995 discovery, averaging more than 30 articles a year from 2005 to 2007, doubling to more than 60 in 2008 and 2009, and holding steady in the 100–120 range from 2010 to 2014.[4] These increases in exoplanet and Mars publications reflect an overall growth in the field of planetary science.

Today, planetary science is an international undertaking. Though my field sites were primarily American and focused on the work of American scientists, international collaborations were common. Not only did scientists from Europe and Asia frequently visit my interlocutors' institutions, but several graduate students and postdocs whom I met at American universities were themselves foreign nationals. Astronomers also frequently travel to facilities in South America, and my own research took me to an observatory in Chile. Investment in planetary science tends to be heavier in nations that have or are developing a space program, but so many data sets are available online that even a researcher or a country without direct access to a satellite or telescope can potentially make a research contribution. Planetary science indeed stretches across our own planet.

Planet as Both a Scientific Object and an Everyday Thing

But what, exactly, is a planet? Is the category—around which I have drawn together my field of study—robust enough to describe objects as diverse as Earth, Mars, and exotic exoplanets? I initially explored this question while puzzling over the circumstances of Pluto's 2006 reclassification from "planet" to "dwarf planet" (Messeri 2010). After an astronomer discovered an object thought to be larger than Pluto orbiting farther from the Sun, the

International Astronomical Union (IAU) moved to create a standardized taxonomy for objects in our solar system. In making explicit what is and is not a planet, the committee in charge of writing the definition necessarily prioritized some scientific interests over others. Astronomers additionally complicated the process of crafting a definition when voicing concern for the public sentiment and the distress some schoolchildren might experience if their favorite planet was suddenly no more (semantically speaking). "Planet" operated as both a scientific and a cultural object.

As defined by the IAU, a planet is a round object orbiting the Sun that is large enough to have either captured nearby debris as satellites or expelled the debris to other orbits. Pluto failed to dominate its orbit and thus was not a planet. This reclassification had little effect on scientific practice. Scientists studying Pluto refer to themselves as planetary scientists (not dwarf planetary scientists). Though exoplanets are not planets by the IAU definition, as they do not orbit the Sun, they are still referred to and understood as planets. Whether studying Pluto, Mars, Jupiter, or exoplanets, scientists use "planet" without hesitation to describe the objects with which they work.

Historian of science Lorraine Daston, in *Biographies of Scientific Objects* (2000), outlines the realist and constructionist approaches to thinking about scientific objects. The realist situates an object as discoverable, as something always existing but not always known. The constructionist, in contrast, depicts objects as inventions; as things molded from a historical and local context. In other words, is a scientific object something that has meaning—that even exists—outside the cultural and historical practices of scientists? A realist answers yes. A constructionist disagrees. Daston offers a middle road: scientific objects are both real and historical. Pluto, for example, was discovered and acquired meaning in 1930, though it surely existed before discoverer Clyde Tombaugh's announcement. As a realist object it has orbited the same path for millennia but was only constructed as a planet until 2006. Whereas one initially studied Pluto to understand more about the icy outer planets, one now looks to Pluto as an example of its neighboring dwarf planets. Scientists adjusted Pluto's ontological status, and in response its epistemological utility shifted.

The reason for Pluto's changed positioning is the result of scientists' reinterpretation of "planet." Planet is the scientific object that I interrogate in this book. I am not offering a Dastonian biography of a scientific object but a series of encounters with different contemporary practices organized around entities that scientists think of as "planets." "Planet" challenges what it means to be a scientific object in several ways. Daston opposes scientific objects to quotidian, everyday objects. Quotidian objects "are the solid, obvious, sharply outlined, in-the-way things. . . . They are all too stable, all too real in the commonsensical meaning of 'hard to make go away.' . . . In contrast to quotidian objects, scientific objects are elusive and hard-won" (2). Planets, however, are both quotidian and scientific. Earth is part of our daily experience, implicated as a planet, thanks to photographs from the Apollo mission.

Yet exoplanets (as opposed to Earth) are more similar to the "elusive and hard-won" scientific object. They are real only insofar as their visualizations are believable (as discussed in chapter 3). Hans-Jörg Rheinberger prefers the term "epistemic things" when considering scientific objects. "Epistemic things" lead the scientist down a path of questioning, as they "embody what one does not yet know" (1997, 28). "Planet" acts like a heuristic, in that knowledge of well-studied planets guides the scientist's understanding of newly detected planets. To label these detections "planets" transforms them from mysterious astronomical objects about which any number of questions could be asked to the focus of a field of inquiry with a well-defined research agenda. The concept of planet also guides research when scientists apply quotidian information about Earth (what a stream looks like) to make sense of an inscrutable scientific object (e.g., a satellite image of a Martian feature). "Planets" hold power as scientific objects or epistemic things precisely because they are at the same time quotidian and scientific.

The final, and most important, way "planets" are unique among scientific objects comes from the experience of *being* on Earth. To inhabit Earth is to be in a place, to move between places, to create and destroy places. Planets, I argue, are more than objects. They are imagined as *places* amenable to habitation (either by humans or other beings). Scientific practice transforms planets from *objects* into *places*, and this movement is an essential way of knowing and doing planetary science.

Planetary Imaginations

To be on Earth is both to dwell in a locality and to be connected to a planetary system. But how is Earth, in all of its vast geography, understood? Imagining Earth as a planet and a globe is not a new phenomenon (see

Cosgrove 2001). However, planetary imaginings seem to be on the rise. Producers of popular culture, for the past several decades, have been telling narratives of our planet that are often focused on global disaster. The silver screen has destroyed or threatened to destroy the Earth through natural disaster, alien invasions, or, as with the 1998 movie *Armageddon* or the more recent 2011 *Melancholia*, a threatening yet impartial extraterrestrial meteor or planet. News stories on climate change, the geographically amorphous "war on terror," and the "global financial crisis" appear on the front pages of newspapers and the homepages of websites, inviting citizens to think on a planetary scale. The Internet itself offers constant connectivity with a network distributed (though not evenly) across the surface of the Earth.

Scholars in the social sciences and humanities have also recently begun thinking about "the planet." Gayatri Spivak (2003) introduces the term "planetarity" as a way to conceive of escape from globalization and its "imposition of the same system of exchange everywhere" (72). Planetarity, perhaps because it appeals to a word associated with "nature" (planet) rather than "culture" (globe),[5] serves to remind us that we are guests of Earth. It is humbling and therefore, one hopes, saving. Similarly, Paul Gilroy in *Postcolonial Melancholia* labels the first part of his book "The Planet" and also chooses to use "the planetary" instead of "the global" as his indicator of scale to suggest a "contingency and movement" that he finds lacking in the concept of "globalization" (2005, xv). Whereas "globalization" suggests an expansive flattening, "the planetary" resurrects a sense of finitude accompanied by the reality of unequally distributed wealth and resources. It is on this uneven terrain that Gilroy situates an "anti-racist" solidarity (75). Yet the planetary simultaneously conjures a solidarity of species—one threatened by environmental and biological offenses that spread agnostically between territories.

Awareness and debate over "planetary crises" (Masco 2010) has only intensified since Spivak and Gilroy wrote of the planetary. Theorizing the planetary is finding ever more traction with the idea of the Anthropocene (see Chakrabarty 2009). Yet the very etymology of the Anthropocene, which scientists have suggested as a name for the current geological epoch in which human impact is noticeable on a planetary scale, risks a reprioritization of the human over the planetary. Bruno Latour (2014, 2015), for one, is wary of this recentering and writes, in a register similar to Spivak and Gilroy, that the Anthropocene should instead enable us to recognize that the Earth is itself an actor with agency.[6] This offers a vantage point from which to think of humans as "planetary subjects rather than global agents" (Spivak 2003, 73). And for Latour, nonhumans are also planetary subjects, rendered on equal footing with humans.[7]

But what are the implications of thinking on the planetary scale? Is it even possible to make sense of dynamics occurring within dimensions so much larger than our daily, human experience? On film, these large-scale dramas are told by focusing on one protagonist (and the protagonist's family). Though there are shots of the impact of the planetary phenomenon on other places and people, the action of these films unfolds in a local, intimate setting. It is difficult to tell a planetary story, and this difficulty is present in the scholarly works I have just cited. Frameworks are offered, but the concepts of planetary and planetarity are left purposely broad and flexible. This is not a weakness of the texts but suggests that analysts are still searching for salient ways to address phenomena of a planetary scale.

I would like to propose that one reason "the planetary" causes us to stumble is because it requires that we grapple with *intangible modes of being*, ways of being that at first blush seem disconnected from *place*. At the same time, place itself is undergoing a change. No longer is place local, but Earth itself becomes a singular locality (a paradox I will return to in the next section). And no longer is place necessarily material; rather it is something that can exist in cyberspace, populated with online worlds. I argue that the planetary scientists who make up the ethnographic center of this book have, out of professional necessity and personal desire, figured out how to cope with and make sense of the planetary scale by reconnecting with the concept of place. How might their invocations of place be useful for other theorists seeking to understand phenomena on a global scale? In other words, what does place *do* and how does it allow us to understand the planetary?

The Earth's surface is almost 200 million square miles. The troposphere, the region of the atmosphere in which weather occurs, extends seven miles above the Earth's surface. The stratosphere extends to 31 miles, followed by the mesosphere and the thermosphere, which climbs to 440 miles. At what point the outermost layer of the atmosphere, the exosphere, becomes "outer space" is a matter still contended. To invoke the planetary is to necessarily wrestle with this expansive geography. As I learned from my scientist interlocutors, place is a powerful tool for pinning down this enormity. It invites familiarity by suggesting that our knowledge of local, recognizable places is possible at a larger scale. For planetary scientists, as I will show, place-making at a planetary scale resists homogeneity. Instead, scientists strive to differentiate features on what, at first, might appear as uniform spheres and describe planets as changing and evolving worlds. They create new kinds of landmarks, like atmospheric "hot spots," that exist on this scale. In so doing, the planetary becomes something that can be navigated, whose dynamics can be observed, and from which lessons can be learned. Place transforms the geographically alien into the familiar.

Creating place also transforms the planetary from the perceived to the experienced. A place-based orientation, rather than passively gazing at the globe from the outside, allows for an imagination of being on/within/ alongside, of experiencing, the planet. This is an active relationship between subject and planet, which for planetary scientists becomes foundational to how they come to know their planet of study. The kind of place they can imagine being on potentially opens up new questions that can be asked about the planet. Experiential ways of knowing become productive avenues for thinking about things on a planetary scale. This is made possible because of place.

When we think of place, we imagine territories that span a few feet, maybe a few blocks, or even a few miles. We do not typically imagine a place many millions of square feet in area. Yet planetary scientists have become quite comfortable with stretching place to encompass the planetary. In so doing, they provide a new interface for considering the planetary, one that has been professionally productive for scientists and perhaps can also be meaningful for social scientists and humanists who similarly grapple with planetary phenomena.

Each chapter of this book teases apart an aspect of what I here call the "planetary imagination." This is not a singular, static, or even robustly definable imaginary but a phrase I use to capture holistic conceptions that scientists have of the planets they study. The planetary imagination includes scientific understandings of the planet and conceptions of planetary pasts and futures, as well as notions of what it would be like to be on and live on other planets. The planetary imagination is yet another way of talking about the placehood of planets, the topic to which I now turn.

Like planetary science, this book brings different disciplines to bear on a central topic. "Place" has been studied in a variety of ways, and most relevant to my theorization is the work of anthropologists, cultural geographers, and scholars in science and technology studies (STS). For anthropology and STS, place is explanatory, used to make sense of cultural expression or productions of scientific facts. Geographers have more often than anthropologists and STS scholars investigated the ontologies of place and place-making. For all three fields, place is a tricky category because it is weighed down by colloquial meanings and associations, making the analytic potencies of the term sometimes hard to parse. The challenge is to liberate place, to question its assumed relationships to other categories and to see how far place can travel — if it can extend into the cosmos.

Cultural geographers have offered distinct and meaningful formulations of "space" and "place" as analytic categories (Tuan 1977; Buttimer and Seamon 1980; Merrifield 1993; Harvey 1993). In the standard formulation, space is universal, empty, and a priori, while places are meaning-filled subsections of space. This distinction is problematic as space, then, becomes associated with the global, the objective, and the masculine, while place is essentialized as a local, feminized subjectivity. In one attempt to curtail these connotations, scholars have deconstructed the space/place dichotomy, doing away with place and advocating for space as the analytic term de rigueur (Soja 1989; Massey 1994). However, like others, I find the term "place" helpful (Relph 1976; Sack 1997; Casey 1993) even as I find the connotations troubling. Rather than excising the term from the literature, I seek instead to destabilize it. Place, as I use it throughout these pages, is not a static and singular term but is multiple and varied, constantly being made and altered (see also Massey 2005b). Place is social (Lefebvre 1974), historical (Harvey 1989; Kern 2003), and political (Zukin 1993; Cresswell 1996). Place is a process (Pred 1984).

To suggest, as I do, that place can be used to describe the planetary scale frees it from associations with the local and in fact makes it a term that can trouble the relationship between local and global in ways that complement Spivak's and Gilroy's intellectual projects. Philosopher Edward Casey (1996) resurrects the importance of place, and its fraught relationship with locality, by uncovering its intellectual grounds. The Enlightenment, Casey argues, subsumed place beneath and behind space, with Kant in particular espousing the primacy of first understanding the general and the global before delving into the local. As the general became tied to space, place was relegated to the local and ultimately understood as a secondary way of knowing.

How, then, to reframe place as something more than local (and thus also as a primary way of knowing)? Casey does so by affirming that spatiality is not something that exists in the mind but is experiential. It is only through experience that one can come to know anything, and experience always begins in a particular place: "There is no knowing or sensing a place except by being in that place, and to be in that place is to be in a position to perceive it" (18). By this he means that the fundamental act of being is an act of being in place. It is from place that all other understandings flow. The global only makes sense because of our knowledge of the local. A planet is comprehensible only once it is understood with the specificity afforded by understanding it as a place.

In anthropology, place has similarly been obfuscated by or made secondary to other categories, most often culture. As Akhil Gupta and James Ferguson (1992) point out in their critique, this has led to almost a century of anthropologists assuming that specific, bounded places are containers for culture. This presumption of place has made it possible for anthropologists to leave "home" and travel to and study "other" cultures, a spatial distancing similar to the temporal distancing Johannes Fabian (1983) describes. Neutralizing (and naturalizing) place made it easy for anthropologists to overlook the interconnections, overlaps, and multiplicities of cultures. Setha Low (2009) suggests that this produces a hesitancy to make place a central concept. Low offers that analyzing the embodiment of place allows for the mobility of place and connects the global with the local (see also Low and Lawrence-Zúñiga 2003). Ethnographies of "the West" and multisited practices also seek to unsettle the notion of static, spatial containers by questioning the categories of us/them and here/there (Marcus 1995; Marcus and Fischer 1999). But in scaling up place to a planetary level, a vantage point from which culture looks comparatively inconsequential, "home" is no longer Cambridge, Massachusetts, but Earth.

To be clear, my project finds the dynamic between "here" and "there" productive, but "there" is a category mutually constructed by the scientists I study and myself. While there is not a geographically bounded community around which my ethnography centers, my interest is in the cultural practices that center on making other planets recognizable as places that can be pointed to, called "theres," and thus studied. While this positioning facilitates a move away from place as a static container, it raises a new set of questions. If other planets are "there" and Earth is "here," who, then, is the "us" who occupy the planet? Planetary scientists speak in a register that collapses all of humanity into "us": "The discovery of [exoplanets like Earth] will change the way we humans view our place in the cosmos" (Lunine et al. 2008, x). Though I will reflect on such cosmological imaginations in the conclusion, it remains beyond the scope of this book to assess the validity and implications of such a universalizing statement. However, such a claim does signal the problems that arise from thinking of place and culture on a planetary scale. The "isomorphism of place, space, and culture" (Gupta and Ferguson 1992, 7) that anthropology has worked to move beyond has a danger of re-presenting itself.

Whereas many scholars in geography and anthropology have thought critically about place, science and technology studies is only just beginning its spatial turn. Attention has been paid to the places of science (Lynch 1991a; Galison and Thompson 1999; Gieryn 2002; Livingstone 2003), but these illuminating studies nonetheless conceptualize place as a location in which scientific practice occurs. Such places are constructed and have important epistemological implications, but place pertains to science only insofar as it describes where science happens.[8] My work asks what other dimensions of place matter for scientific practice and concludes that understanding how scientific objects themselves are constructed and understood as places points to a spatial logic at work beyond places of science (e.g., the laboratory and the field). Along with other programmatic calls to reevaluate how history of science and STS think about place (Powell 2007; Shapin 1991, 1998), I offer a direction for the spatial turn that expands how placemaking figures into scientific work. This follows the work of the geographers who have freed place from being subsumed by space and the anthropologists who have severed the connection between place and culture, as I seek to offer a line of inquiry that liberates place from exclusive associations with *where* science happens. Place should be examined not only as a location for practice but also as a resource scientists use to create their scientific objects as such. Place is an epistemological heuristic; a way of understanding that is actively pursued and cultivated by, in this case, planetary scientists that allows for a meaningful mode of interacting with objects that physically lie outside human experience. Place is not an afterthought or something produced alongside science but is intimately tied to daily practice. Producing science is producing place.

The Sociality of Outer Space

Following the lead of planetary scientists, a small field that can be called the "social studies of outer space" similarly seeks to understand what the cosmos can tell us about ourselves. This book contributes to this growing set of scholarship that probes the cultural connectivities between cosmic worlds.[9] Peter Redfield's *Space in the Tropics*, a history and ethnography of French Guiana, is an especially noteworthy entry in this literature. Redfield, whom I quote in the epigraph, motivates his study by noting the geographical juxtaposition of two seemingly incongruous projects: France's high-modernist Ariane rocket project of the twentieth century and the colonial project of the penal colony of the prior century. He opens his book by asking about the role of place: "Does it matter where things happen? Or more precisely, what might it reveal that different things happen in the same place?" (2000, xiv). Redfield describes how the project of space exploration has produced "placeless space," a descriptor both for the connected information landscape brought about by communication satellites and for the control center of the Ariane launch facility that obfuscates ties to local cultures and histories (a placelessness that I revisit in chapter 4 when discussing mountaintop observatories). Redfield writes at the end of the twentieth century, a century that launched humans into space and allowed all to see that "what lay beyond [Earth had] become meaningful as space, a vast sea of darkness surrounding a blue and green point of human place" (123). The space exploration triumphs that I catalogue at the beginning of the twenty-first century are marked by robot, rather than human, travelers and involve a deepening understanding of the planets in our solar system and in others. A new geography is supplanting "placeless space." Planetary places are populating the "pale blue dot" cosmology as Earth becomes part of a vast interplanetary network.

To begin figuring out how to make sense of this new and emerging configuration, David Valentine, Valerie Olson, and Debbora Battaglia have made a plea for anthropological studies of outer space, arguing that "outer space is a crucial site for examining practices of future imagining in social terms, and for anthropological engagement with these practices" (2009, 11). Space, they offer, allows "theorists of space and time" a venue on which to test the limits of their theory. To operationalize studies beyond Earth's atmosphere, they put forth the concept of the "extreme" (Valentine, Olson, and Battaglia 2012; see also Olson 2010). The extreme is definitionally expansive, meant to put into conversation the ordinary and extraordinary, the past and future, the worldly and other-worldly. Stefan Helmreich's (2009) ethnography of microbiologists who study extremophiles (organisms that live in extreme environments like thermal vents on the ocean floor) has shown how the study of these extreme microbes inspire scientists to make comparisons between places on Earth and the environments of other planets. The locations where the extremophiles thrive are uninhabitable by humans, raising questions of how life might survive on what humans think of as uninhabitable worlds elsewhere in the solar system and beyond.

These ethnographers challenge other scholars to understand "'space' as a site that is always already occupied by humans, human knowledge production, and human imagination" (Valentine, Olson, and Battaglia 2012, 1010). Outer space is social, and studying it as such will help articulate the changing dynamics between our world and others. Drawing inspiration from planetary scientists, we can reconceptualize the universe not as a void but as densely inhabited by planets and other cosmic objects that tell us about Earth. Whereas scientists make comparisons and draw analogies between planets to better understand the physicality of Earth and its geologic past and future, anthropologists can follow this "comparative planetology" approach and apply lessons learned from studying the sociality of outer space to make sense of what it means to be on Earth. This is especially important in an age of "planetarity."

If we allow that outer space is a site of sociality, we must also question what social relations are or will be reproduced beyond Earth. Theorists of place warn of the inherent hierarchies and exclusions that come with place-making practices (Cresswell 1996). If, as I argue, place-making is, for scientists, a primary mode of relating to outer space, what difference-making practices are also taking flight? The difficulty with answering this question is that space science and indeed many space endeavors are wrapped up in a utopic narrative. Studying the cosmos or traveling to a planetary destination is, simply put, understood as being for the good of humankind. Valentine's ethnography of NewSpace, companies and communities that work toward commercializing space flight, reveals that a widely shared goal is, surprisingly, orthogonal to profit. NewSpace entrepreneurs hope to do what NASA could not: settle outer space. Capitalist and libertarian beliefs (and accompanying inequities) are of course tied up in this enterprise, but Valentine wonders what studying NewSpace looks like if, rather than focusing on the replication of markets and profits into space, one takes NewSpace entrepreneurs seriously in their claim that ultimately NewSpace is humanity's hope for a future beyond Earth. This requires postponing the search for replicated differences and instead engaging "with contemporary human activity that is not already explained by the brief span of modern human history" (2012, 1063). A dizzying endeavor to be sure.

It is this orientation that I often adopt when parsing the rhetoric I most frequently encountered in my fieldwork. Rather than "settlement," the NASA and university scientists I worked with framed their work within the aligned, place-centric project of "exploration." Humans have "explored" the Moon, and robots are "exploring" Mars. Exoplanet astronomers dream of "exploring" planets around other stars. With the exception of the Moon, this exploration is remote, yet there is a pervasive imagination of human, emplaced exploration. The rhetoric of exploration is powerful and has a long, sometimes damning history. The aim of this book could have been to unpack the white, American, imperial subtext of invocations of exploration. How does this rhetoric extend the colonial narrative beyond the postcolonial Earth and into space? It is interesting to imagine what such a reading of this particular sociotechnical network would look like, when the stakes of empire building are not about the physical acquisition of land, nor are the territories being explored peopled (that we know of). When appropriate, I do point to how exploration, as planetary scientists talk about it, contains references to other projects of state, nation, and empire building and how who can be an explorer or have access to tools of exploration creates hierarchies and differences (see specifically chapter 2). But, like Valentine, I have most often found it prudent to understand exploration as my interlocutors do. Though I attempted in several interviews and conversations to question scientists when they threw out "exploration" as the obvious justification for their work, this line of inquiry was fruitless. Exploration is an unquestioned good for this group of planetary scientists. Thus, rather than trying to demystify or debunk this constant invocation, the work of this book is to take that conviction seriously and understand how it invites a changing sense of place in the cosmos.

Star Chart

How, then, to describe the work of place-making that I claim scientists perform regularly? I observed several different activities, or techniques, of place-making at my different field sites. These activities guide, in fact give name to, each chapter. Scientists rely on narrating, mapping, visualizing, and inhabiting to imagine themselves on other worlds. Narrating builds a rich story that connects Earth with another world. Like the story that exoplanet astronomers tell about a future in which one can point to the sky and identify a world like our own, narrations reach across both space and time. Mapping and visualizing other planets translates the strange and unknown into the sensorially relatable. Planetary scientists push their practice to make visible the invisible. They want to be able to see the curvature of a dried-up streambed on Mars or the glow of an exoplanet. Whereas decades of optical development makes it possible to photograph the streambed on Mars and create incredibly high-resolution maps, exoplanet astronomers still study mostly unseen planets. Astronomers have been able to directly image only a few exoplanets, and these portraits appear no different from a star; they are points of light with unresolvable surfaces. Exoplanet astronomers must develop novel ways to "see" exoplanets by experimenting with nonpictorial representations. These visualizations nonetheless function like maps as they become a way for researchers to understand the exoplanet as a place and navigate its surface.

Narrating, mapping, and visualizing rely on spoken and visual language to conjure another world. Inhabiting and forms of embodiment are tools of place-making employed even when the place being made is physically inaccessible. When I speak of inhabiting, though, I am suggesting less a technique of place-making and more an aspiration for place. How planetary scientists inhabit the locations in which they work, their labs and their fields, certainly influences their place-making. But the verb "inhabiting" signals the scientists who aspire to find places of habitation beyond the Earth. "Habitability" is a term planetary scientists use to conjure images of a particular kind of place. I use "inhabiting" to connect emplaced practice on Earth with the imagination of habitable planets. Planetary scientists make planets into places by combining these four techniques of place-making in different ways. In each chapter different techniques will complement each other, but I will keep analytic focus on one dominant mode of place-making that best characterizes the work of each of the communities I encountered.

This book proceeds as a journey, beginning here on Earth before traveling to Mars and then on to exoplanets orbiting other stars in our galaxy. By the end, however, we will find ourselves right back where we began, craving to connect with our own planet. The interplanetary itinerary begins with chapter 1, in which I show how planetary place is made right here on Earth. Through activities at the simulated Mars habitat of the Mars Desert Research Station (MDRS), I consider how Earth itself is transformed into a Martian place. In superimposing ideas of outer space on the terrestrial landscape, the site becomes multiple, simultaneously Earth and Mars. Narrative both helps uncover the dynamism of these landscapes and connects present "analog" work to earlier forays into planetary superposition both in the field of geology and during Apollo astronaut training. Past and present currents join speculative future imaginings of life on the Red Planet to structure one's experience at the MDRS and consequent understanding of what it might be like to be on Mars.

In chapter 2 I discuss my participant observation at NASA Ames Research Center in Silicon Valley and the work of creating the highest resolution 3-D maps of Mars yet produced. Attending to these maps and how they are made demonstrates that NASA scientists portray Mars as dynamic, immersive, and accessible to all. I more deeply examine the rhetoric of "exploration," used ubiquitously at NASA to describe aspirations for Mars, and questions of whose vision of exploration is here represented. How exploration is discussed and Mars is depicted are different today than they have been in the past and are a product of institutional currents both at NASA and in Silicon Valley's startup culture. Situating this extraterrestrial mapmaking within its specific location further demonstrates that planetary science is not a niche or isolated science but very much embedded in contemporary conversations and activities.

Chapter 3 leaves the solar system, following the exoplanetary work of Sara Seager and her students and colleagues at MIT.[10] The first exoplanets found were objects larger than Jupiter, orbiting extremely close to their host stars. These planets are too far away and too hostile to accommodate human presence, yet during my fieldwork I repeatedly encountered astronomers poring over graphs and models of individual planets, imagining the weather and surfaces they might host; imagining what kind of places they were. Following the visual and semiotic practices of this scientific community-in-the-making offers a case study in pedagogy and how norms of seeing are developed and taught. Learning to be a successful exoplanet astronomer is learning to see worlds in data and disciplining others to see in the same way.

Finally, in chapter 4 I consider the search for "Earth-like" exoplanets. This chapter draws on my experience with astronomer Debra Fischer during a trip to an observatory in search of planets around the Sun's closest neighboring star, Alpha Centauri. What, I ask, is the role of the observatory in the increasingly remote and automated practice of observational astronomy? This question is accentuated when considering the search for planets like our own. Astronomers present these elusive Earth twins as "habitable" planets: rocky planets, similar in size to Earth, that are at a distance from their star such that liquid water can exist on the surface. At the same moment that they are hunting for habitability elsewhere they are decreasingly inhabiting their place of work, shifting centers of practice away from ground-based observatories and toward computers and virtual databases that can access satellite telescopes from anywhere. Yet, as I show in this chapter, even as virtual presence replaces the need to physically be at an observatory, on the rise is a desire to find other home-worlds in the cosmos and other places humans can be. The search for other Earths contains a longing to (re)connect with our own planet.

Alongside this examination of place, I develop the idea of the planetary imagination throughout the book. In the first chapter I establish the aspirational aspect of the imagination; how it captures beliefs and hopes from the past, present, and future of what planets *are* and thus what they would be like to occupy. The planetary imagination is a shared imaginary among the scientists among whom I circulated, but they also desire to share the planetary imagination with a broader audience. Chapter 2, then, examines how maps of Mars are tools for this circulation. Mapmakers embed aspects of the imagination, aspects that were also present at the MDRS, in these maps in a way that enables lay users to access a similar imagination of being on Mars. The rich planetary imagination that scientists associate with Mars is stretching deeper into the galaxy as exoplanet astronomers engage with and develop it further. Having already addressed how the imagination circulates to a wider audience, I elucidate in chapter 3 how students and new members to the field are trained to see this imagination in the science they are producing. Finally, the search for an Earth-like planet ties the planetary imagination that is reaching deep into the cosmos back to our own planet. Earth itself, it seems, is the ideal type of the planetary imagination.

But *why* are, as I claim, scientists keen to populate the universe with infinite worlds on which they can imagine being? I address this question in the conclusion, tying the proliferation of planetary imaginations, and the question whether Earth is unique, to the current cultural moment. Speculating on the existence of other worlds has inspired theological treatises, has been a way to debate the relationship between religion and science, and as recently as the mid-twentieth century has mediated the anxiety of becoming a space-faring species. Whereas traces of these impulses persist, what dominates the asking of this question today for the community I studied is a longing for connection. That longing is captured in the image with which this book opened: a person pointing toward a star and understanding Earth's relation to other worlds. Conceiving of planets as worlds and determining how many there are and if any of them are like our own situates Earth in a larger cosmic context that connects work and life here to being elsewhere. "Knowing our place in the universe" is knowing the planetary network in which Earth exists.

The ways planetary scientists construct planets as places suggest that *place* is a mode of understanding. The act of "placing outer space" puts forward not a particular worldview but a cosmosview. Some cosmologists imagine the structure of the universe to resemble a cotton ball pulled apart. Wisps of material form a thin web of cotton, with denser nodes dotting the surface. These nodes, on a universe scale, are "superclusters" of galaxies, pulled together by gravity. Each supercluster itself resembles the structure of the pulled-apart cotton. Galactic clusters form the nodes in the supercluster. The Milky Way is thought to be in a cluster of thirty other galaxies, our Local Group. Each galaxy, in turn, is filled with massive and mysterious objects like nebulae and black holes.

The cosmosview of planetary science sets aside these structures, each more mammoth than the previous, and concerns itself with tiny globes of gas and rock that orbit relatively small, stable stars. This proverbial search for and study of needles in haystacks is rich with meaning and implication. These objects are just barely on the scale of our comprehension, yet scientists frame these planets as entities that bear on our own planetary existence. They enrich the connection between *being* and *planets* by offering evidence of the placehood of these objects. In so doing, planetary scientists are populating the "emptiness" of space with planetary places. Thus, they are offering a new vision of the universe; they are replacing outer space with outer place.

CHAPTER ONE

N A R R A T I N G   M A R S

I N   U T A H ' S   D E S E R T

The Mars Desert Research Station is located outside the small town of Hanksville, Utah. At this facility, run by the Mars Society, participants engage in two-week simulations of living and working on an early Martian settlement. The closest airports—Grand Junction, Colorado, and Salt Lake City, Utah—are both several hours away from the site. At two-week intervals, from November through May, six people—often strangers, sometimes colleagues—meet up and drive from the airport to Hanksville. During this drive, these "crews" are treated to the vistas of the American West. Grey, purple, and red rock formations hug the highway; McDonald's drive-thrus give way to buttes; strip malls to canyons. As the crews draw closer to MDRS, they hope that another shift will occur: that Utah will become Mars.

Once at MDRS, crew members will often reflect on their arrival in the first report they submit to "mission control," an offsite team of volunteers responsible for monitoring these daily reports and posting them online.[1] Consider one such account of a visitor making the journey from the airport to MDRS and how she tells the story of her crew's journey from civilization to isolation; from Earth to Mars: "The journey from Grand Junction, Colorado to Hanksville, Utah in itself was beautiful. The transition from the norm to alien terrain was really quite breath-taking. . . . We headed out to the Desert Station. . . . The Sun was setting over the nearby hills, hugging every last red, orange, and brown rock goodnight. The dance between light and shade over the landscape and the Station was ever-changing and beautiful. It really was an [sic] stark outpost lost in the ocean that was a desolate but mesmerizing world" (Dale 2011). On viewing the MDRS living habitat, or "hab," and its environs for the first time, other participants have gone even further in transforming the landscape from Earth to Mars: "This beautiful landscape is a glimpse of what a real Martian landscape might look like. It felt more than just a setting for a Mars simulation. A more accurate feeling: this IS Mars!" (Gough 2002). These are narratives of arrival, of situating oneself within a mental and physical state from which to transform one planet, one place, into another. Narrative constructs the landscapes surrounding MDRS as Mars, Earth, or a combination of the two. In spinning stories about the environment—labeling the landscape "alien," for example—visitors to MDRS carve out a unique place to inhabit and consequently forge a novel connection to or understanding of another world. The imagination employed by visitors to MDRS superimposes the planetary on the local. This requires reconciling a vast difference in scale. This chapter will illustrate how narrative bridges this scale, brings the planetary into the realm of place, and thus makes the planetary in general and Mars in specific something that can be embodied, experienced, and thus better known.

Cultivating Mars on Earth

How did Mars come to exist on Earth? One of the actors closely involved with crafting the ethos that fuels MDRS is Carol Stoker, a planetary scientist whom I met at NASA Ames in July 2009 during a preliminary fieldwork trip that led to the participant observation I describe in chapter 2. Following this trip Stoker and I stayed in touch, and she subsequently invited me to join her at MDRS. At our initial meeting, Stoker welcomed me into her office to talk about her research projects. Stoker is a slender woman with long, slightly wavy, light ginger hair. She wears large glasses that accentuate her inquisitive eyes. She was born in Utah and has a certain brusqueness of personality and a hint of a drawl that speaks of being raised in an environment shaped by a reverence for Butch Cassidy's Wild Bunch. At eighteen, Stoker left home and hitchhiked to California. She intended to imbibe the 1960s counterculture of Haight-Ashbury, but the fever was already subsiding. She returned to Utah for a bachelor's degree and went on to earn a

PhD in astrogeophysics from the University of Colorado, Boulder, in 1983. When she arrived at CU-Boulder, Stoker quickly made friends with fellow students who, like her, were inspired by the Viking lander images of Mars and were disappointed by NASA's inability to mount a human mission to the Red Planet. After years of plotting and thinking about exploring Mars among themselves, they convened a conference in 1981 called "The Case for Mars." At this conference the Mars Underground—a group that grew out of conversations between Stoker and her classmates—officially announced and established itself.[2]

Over the next decade and a half, they continued to hold Case for Mars conferences every few years. The founding members of the Underground moved away from Colorado, and Stoker landed at NASA Ames in 1985 and has been there ever since. When I visited, her office was decorated with Mars memorabilia. A Looney Tunes Marvin the Martian figurine looked down from a bookshelf, and a flag depicting Mars, designed during the Mars Underground's heyday, hung above her desk. Stoker explained that much of her research is "analog research." Planetary scientists travel to places on Earth they think of as "Mars-like"—analogous to another planet's terrain—to test out equipment or do geological research. She suggested that these sites have been cultivated in the past twenty or so years out of frustration by people who would rather be using equipment and doing geology on Mars. Stoker freely admitted that she was and still is one of those people.

Stoker's research at analog sites involves the testing of drilling equipment that might be used on Mars. Her two primary analog sites of recent years are Rio Tinto, a river in southwest Spain, and MDRS. During an MDRS field season crews of six apply for two-week access to the facility. It differs from Rio Tinto, she explained, because it is a *simulation*. Not only do people go there to test out equipment (as it is one of the best analog terrains of Mars that Stoker says she has seen) but also they can engage in practices that establish the habitat as "safe" and the outside as "hostile." In Rio Tinto, the team might stay in motels or tents. At MDRS, the cylindrical habitat is outfitted as its designers imagined an early settlement on Mars might be. There is also an implied obligation to wear simulated space suits whenever leaving the habitat, as the rules of the simulation stipulate that the "hab" is surrounded by the unbreathable Martian atmosphere (see fig. 1.1). For many crews, this is a crucial element of the simulation. As reported by one participant in a daily dispatch: "The crew consensus was that before

1.1 Mars Desert Research Station participants in full simulation returning to the habitat. Photo by the author.

you suit up in the 'space suit' we are simply roughing it and trying to do some science on the side. Once you put on the suit and are surrounded by an otherworldly scenery, everything becomes surreal.... you are on another planet" (Gale 2005a).

Stoker was planning a few missions to MDRS in the upcoming months with two goals: first, to test the latest drill her engineers were building, and second, to make the argument to NASA that this site could produce good science and the agency should support it. At the end of the meeting I asked if it would be possible to join one of her crews. She said she would keep me in mind should a spot open up.

MDRS is the second analog station established by the Mars Society, a nonprofit organization that advocates for human exploration and settlement of Mars. Robert Zubrin, who began attending the Case for Mars conferences in the late 1980s, established the Mars Society in 1998, two years after the sixth, and last, conference.[3] Shortly after its founding, the Mars Society entered a partnership with NASA to establish the first analog site, Flashline Mars Arctic Research Station, located in the Arctic on Devon Island. Stoker was there to help raise the roof, literally, for the habitat. However, after one season of joint operation, the Mars Society's goals of public relations and grassroots advocacy created tension with NASA's focus on science and research (Fox 2006, 43). The two organizations agreed to part ways.[4] During the summer field season, the Mars Society continued to simulate missions from the habitat, while NASA crews established a tent city some kilometers away. Several NASA employees remained on good terms with the Mars Society, Stoker included, continued to do research at Flashline, and were equally enthusiastic about Zubrin's plan for a second station as they had been for the first.

Zubrin narrates how the site for the desert analog station was chosen in his memoir, *Mars on Earth* (2004). He became acquainted with filmmaker James Cameron when Cameron spoke at the Second International Convention of the Mars Society in 1999 about his planning for a 3-D IMAX Mars movie. Knowing that Zubrin was in search of a Mars analog site, Cameron suggested some regions his crew had scouted during preproduction. In particular they were very excited about a location near the San Rafael Swell in Utah. As Cameron's search coordinator exclaimed to Zubrin, "It's Mars!" (159). Following up on this lead, Zubrin and the Mars Society decided in the spring of 2001 that MDRS would be built in a red Jurassic desert a few miles down the road from Hanksville. Zubrin summarized his vision for MDRS in a written report during his stay at MDRS in 2002 as part of the inaugural crew: "What we will attempt to do is conduct a sustained program of field research into the geology, paleontology, microbiology, etc, of the area while working in the same style and under many of the same constraints as humans will have to do when they explore Mars. By attempting to produce the maximum science return we can while operating under Mars mission type constraints, we hope to start learning how to effectively explore on Mars" (R. Zubrin 2002).

In late September 2009, Stoker emailed me asking if I was still interested in accompanying her to MDRS. I was immediately brought up to speed on the project through a series of teleconferences. We would be testing out a prototype of a drill she named the Mars Underground Mole (a reference to the organization she started in graduate school many years prior). The November mission was the first of a multiyear project. As such, we would be charged with scouting future drill sites. Stoker, Devon (a NASA engineer), Julia (an astrobiology graduate student affiliated with the European

Space Agency), and I would be at MDRS the whole time.[5] Jon (a NASA engineer) and Natalie (a NASA geologist) would be there for the first few days and then replaced by Danny (a student who interned with Stoker) and Brian (a NASA project manager). In addition to site selection, engineering work, and astrobiology experiments, Stoker wanted to document crew life for ongoing human factors research and charged me with tracking the kinds of tasks with which people filled their days.[6] What routines from this two-week simulation might be relevant when planning for a real human mission?

Double Exposures:
Place and Time, Landscape and Narrative

While living and working at MDRS, I encountered several different place-making practices. As we walked across the desert, we created informal maps and marked GPS waypoints to remind us of spots that were particularly Mars-like. Visualizations, particularly geological ones, were often necessary for figuring out where we stood in place and, as I will describe, in time. And inhabiting MDRS and coping with its infrastructural hardships that altered how we washed and cooked made it easier to "play" the part of Martian pioneers. Though I will reference these techniques throughout this chapter, in subsequent chapters I will take up each of these modes of place-making in turn and examine how they manifest at other sites of planetary scientific practice. But MDRS is fundamentally different from the other research sites I will discuss because the planetary object of study is no longer observable only from a distance, as astronomers are accustomed to. Rather, the promise of MDRS is to offer the bodily experience of studying Mars up close while never leaving Earth.

At MDRS, the landscape is one of the most important triggers for transforming Earth into Mars: "Stepping out of my dark cabin, I immediately found myself face-to-face with the main porthole of the upper deck, and a red-tinted brown landscape of rock and sand stretching out onto the horizon. That's when it really sank in: for all intents and purposes, we were on Mars" (Ruff 2012). This is an example of what I call a "double exposure" of place. But such a double exposure, just as in film photography, is a juxtaposition of time as well as place; a juxtaposition of two snapshots taken under different circumstances. For this visitor, a future that includes the human exploration of Mars is exposed atop the present simulation. When encountered accidentally, double exposures can be jarring, though perhaps welcome, surprises. Consequently, the double exposure of Earth and Mars, present and future, is what makes one's experience at MDRS thrilling: "How grand to own a home in this place! I glance over at the Mars Desert Research Station, straight bright-white and tall. . . . Yes we are here, in this place and in our minds, Mars on Earth" (Clancey 2002). In figures 1.2a and b I invite the reader to imagine a Mars/Earth double exposure, noting how the landscapes in these analog sites are simultaneously comprehended as earthly and otherworldly.

How, then, to make sense of these landscapes infused with multiple places and times; how to make sense of the kind of planetary place being here made? My analysis is informed by interpretive approaches taken by cultural geographers, which equally account for the material and the social when seeking to understand landscaped places.[7] These framings productively figure the multiplicity of place, but to capture also the temporal layer inherent in the double exposure, I offer narrative as a device that structures both place and time as they manifest in landscape.[8] To wit, multiple exposures overlay temporalities alongside spatialities, and narrative helps put these discordances in order. One planetary geologist described to me the appeal of geology, suggesting a narrative quality. "I like the story of how something came from something to what it is today. They all fit into some kind of temporal line."

A narrative approach unsettles landscapes as static images. Eric Hirsch (1995) has described how the pursuit of dynamism distinguishes the anthropological approach to landscape analysis from the geographical approach. This view of landscapes as processes is evident in Hugh Raffles's history and ethnography of the Amazon River. He captures the liveliness of this South American landscape with his phrase "in the flow of becoming" and offers a "biographical landscape," suggesting that how people shaped the river informs our imagination of it today (2002, 5).[9] As with a narrative, the past unfolds into the present. Keith Basso develops this connection between place and time: "instances of place-making consist in an adventitious fleshing out of historical material that culminates in a posited state of affairs, a particular universe of objects and events — in short, a *place-world* — wherein portions of the past are brought into being" (1996a, 6). At MDRS, participants construct place-worlds not only by invoking the past but also through references to the present and future of both Earth and Mars.

1.2 Double exposure of place: one can imagine these two images, (a, top) from Utah and (b, bottom) from Mars, superimposed on one another. Image credits: (a) author; (b) NASA/JPL.

As the phrase "place-world" implies, whole worlds can be brought into being at a defined locality. At MDRS, it is an entire planet that finds materiality through the landscapes and ordering narratives woven by participants. A planetary imagination is enacted as Earth becomes Mars, and double exposures are stumbled upon and made sense of. In the rest of this chapter I will present four static snapshots of landscapes that in fact contain dynamic, complex double exposures of place and time. I call the four stories that order these landscapes the geological, astrogeological, areological, and science fiction narratives. The geologist who accompanied us for the first several days of the mission helped us understand the *geological narrative* of the MDRS environs. The geological narrative is pieced together by decoding clues embedded in present rock formations to make sense of the past. In the 1950s, well before MDRS was built, some geologists wished to extend the geological narrative beyond Earth. Geology, they argued, could be a way of knowing not only on Earth but also on the Moon and other planets. Though this was not a move favorable to the entire discipline (as I will discuss), support from the Apollo program allowed for the weaving of the first threads of an *astrogeological narrative*. The astrogeological narrative is a way to understand not the past of this planet but the present of a distant planet. For the Apollo mission, astrogeologists trained astronauts in geology. Instead of focusing on the terrestrial story of a rock, they taught astronauts to decipher a story that would connect a rock on the Moon to a rock on Earth. Following Apollo, when viewing satellite imagery from Mariner, Viking, and subsequent missions, planetary scientists spun astrogeological narratives about rocks on Mars.

Whereas the first astrogeologists were trained in traditional geology programs, those studying Mars today are just as likely to come from a planetary science background. Often they are more familiar with the geology of Mars, the study of which is called areology, than the geology of Earth. They do "fieldwork" using images from satellite and robotic missions. When they go into the terrestrial field, it is in search of analog sites deemed similar to Mars. Consequently, during such fieldwork they synthesize an *areological narrative*. This reverses the astrogeological narrative, in which another planet's present is understood through the lens of Earth, such that Earth's present is understood through the lens of Mars. With the areological narrative, a planetary scientist makes sense of a terrestrial place through a familiarity with formations and processes on Mars that they have studied.

The geological narrative brings the past into the present, and the astro-geological and areological narratives invoke the presents of different locations. At MDRS specifically, there is also a rich *science fiction narrative* at work. The MDRS facility is imbued with elements from speculative musings concerning a future habitation on Mars. This environment invites participants, at times, to pretend they are on Mars. In entering a science fiction narrative, participants bring elements of the future—of another place's future—into the present. I conclude this chapter by suggesting that an overarching narrative, a utopian one, organizes the multiple narratives, imaginations, and places that come to be at MDRS.

The four landscapes and their accompanying narratives weave together past, present, and future imaginations of Mars. Throughout the chapter I show how scientists redefine what it means to be on Earth. Under what circumstances does being on Earth become proxy for being on another world? A planetary imagination shapes local, embodied experience such that one can claim to know, perhaps even to have visited, another planet.

Landscape 1:

The River in the Rock Bed; A Geological Narrative

We began mission day 2 hunting for a drill site that, based on the Google Earth satellite imagery Stoker studied prior to our mission, looked promising. After a comparison of maps, none entirely reliable, we thought we knew how to get to the point of interest. With GPS and paper maps in hand, we loaded into a four-wheel-drive truck, and Stoker, acting the part of the commander, joked that we were "launching at 10 A.M." after a successful "preflight check." The ignition was fired, and we headed down the bumpy desert road. At some places the road was marked by rows of stones on either side. In other places, there were only tire tracks to suggest we were following the designated path. After a rainfall the road blends into the rest of the surrounding landscape. We took a guess at where to turn off the main road to find our destination. At first the dirt road was similar to the one we left, but after some twists and turns, the road became more groomed, and it was apparent we had made a wrong turn.

We did not find the site from the satellite picture, but we did find a quarry, accessible both by these back roads and the main paved road that

1.3 NASA geologist Natalie stands in front of an outcropping atop a rock quarry outside Hanksville, Utah. She is drawing the layers in her field note-book. Note that to the right of Natalie are varying layers in the bottom quarter of the outcrop and to the left are more homogenous layers. Note also the discolored spot in the middle of the outcrop. Photo by the author.

connected us to the nearby town. We stepped out of the car to explore. Because it was a rock quarry the cliff face was exposed, and we were offered a privileged view of the geologic strata in this area. Geologist Natalie's eyes lit up, and she began constructing a story to explain the exposed layers of the rock face. After taking in the whole formation, she guided me to a specific point on the rock, shown in what I call Landscape 1 (fig. 1.3). The rock was layered in red and gray. Natalie explained that this was silt and sand stratification, which was evidence of a rapidly changing environment. As she followed the strata along the face of the wall, suddenly, in this Utah desert, she saw the signs of an old river. She stood in front of the wall with her arms wide open, imploring me to see how the silt suddenly subsides. Thanks to this cross section, we saw that the gap in silt was refilled with newer rock. This means, she explained, that a stream used to run through this place. Using other geological clues, Natalie told me that not only did water flow here, but it was a raging river, full of energy. As she reconstructed what this place used to be, the dry desert slipped away, and in its place was the damp swamp or forest that filled this area 150 million years ago, before plate tectonics moved its position further and further away from the equator.

Through a geological narrative, the site where we stood became multiple. It was no longer merely a rock quarry but an exotic place where the rocks spoke of mysteries long since past. Natalie, in telling a story that connected the clues hinting at past riverscapes, gave order to the chaotic rock formation. In the middle of the immaterial, invisible river, she spotted a discolored area a few meters up. She began spinning tales of what this discoloration might be. Her final proposal was that it was an island. In the middle of the river, she suggested, a small island had formed and the discoloration might be evidence of the gathering of organic material. We did not have time for a closer inspection, however. Stoker, frustrated that we could not find the site we were looking for, decided to enlist some local help in the form of the resident geologist of the Bureau of Land Management.

Buzz, a name with a fitting space-age ring to it, greeted us at the Bureau of Land Management with a look suitable for a rancher or cowboy. Tall and slender, he wore jeans and a plaid shirt, accessorized with aviator sunglasses and a white cowboy hat. He was soft and slowly spoken but with a sharp sense of humor. Stoker explained to Buzz the drill sites we were looking for, emphasizing the geologic formations to which she hoped to gain access. She was careful to stress the NASA affiliation and assured Buzz that we would do no ecological damage, clearly nervous that he would deny us land access. Buzz did not take much convincing, and after Stoker's fifteen-minute rapid-fire justification he responded with a simple drawl, "I think I might be able to accommodate you."

Buzz first led us back to the quarry in order to show us the difference between the Morrison and Summerville Formations. Natalie was interested in discussing the island she found in the rock wall. Buzz stepped right up to the discoloration and took out a knife to chip away at it. As a piece crumbled off in his hand, he declared that it was actually petrified wood. The imagined island disappeared and was replaced by a forest that flourished many thousands of years after the river had dried up. The petrified tree was no less fantastical than the island. Natalie sighed in amazement that after 140 million years, one could still see the structure of the tree trunk.

This episode recalls an instance written about by anthropologist Fred Myers's in his account of Pintupi place-making in the Western Desert of Australia. In different deserts at different times, he and I both witnessed "how landscape is assimilated to narrative structure" (Myers 1991, 64). Myers recounts a time when he was with a group of Pintupi men who found an oddly colored rock in the landscape that they could not identify. They called on the elders—who had close relationships with the traditional owners of the area and thus possessed knowledge of the land's legends. At the site, "the men chipped off a bit and examined the colors, then dug around the area to expose more of the rock" (64). An elder proclaimed the rock to be part of the Kangaroo Dreaming, the local myth that connects land, people, and their history. According to the dreaming, two men had speared a kangaroo five miles away. Here, the elder speculated, must be where they had gutted it. The discoloration of the rock was the result of the kangaroo's stomach contents.

There is an interesting parallel between these two acts of narration. In both cases, experts were called forth to illuminate a particular rock formation. For the Pintupi, the discoverers themselves were unable to make sense of the discoloration, but the elders used their knowledge to incorporate this new finding into the local cosmology of the Kangaroo Dreaming. While Natalie had a theory about the discoloration, she readily acknowledged that she lacked the necessary local knowledge to understand fully the geological narrative before which she stood. Buzz, the elder we consulted, presented an alternative tale and because Natalie had not previously inspected "the island," she was easily swayed by his local expertise. In both cases an anomaly was swiftly folded into a story that ordered the land and its past.

For Pintupi and geologists "the landscape itself offers clues about what may have happened. Not only does it reveal something about the invisible, but it offers a link to the invisible forces that created it and whose essence is embodied in it" (Myers 1991, 67). What is invisible to most people when gazing at a rock formation but is revealed through the geologist's narrative is the landscaped past: the geological history of the place where one stands. For the scientists I stood with in Utah, the invisible forces were those of plate tectonics, powerful water channels, and sweeping winds.

To uncover the invisible is to learn how to see. In the early days of a field, scientists experiment with a range of visualizations, slowly codifying what Martin Rudwick (1976) would term a visual language. Rudwick developed this concept in his historical study of geology. He argued that the proliferation of pictorial representations arrived at a time when geology was establishing itself as a discipline at the turn of the nineteenth century. With new modes of representation came new modes of seeing. Becoming a geologist became inextricably linked to a certain way of seeing. Geologists learn to see rock formations as historical narratives. Natalie took it on herself to teach the rest of us to see as she does — as a geologist does.

There were several different techniques by which she taught us to make sense of the landscape. My first lesson came during the drive from Grand Junction, Colorado, to Hanksville, where we would formally begin our mission at MDRS. Natalie frequently gestured to the stark landscape, describing how to recognize the most prominent geological formations. The Morrison Formation, which spans from northern Arizona to southern Canada, is lightly striated, ranging in color from ash grey to a rusty red. Formations, she explained, are named for the first place they were studied. This formation was first documented in Morrison, Colorado. We were well into Utah by this point in our drive, but the landscape created a geological connection between where we presently were and the place we just left. As we drove, Natalie apologized when she hesitated before naming a formation. She had meant to brush up on this area before the trip. When we arrived at Hanksville, we stopped in a convenience store named Hollow Mountain, where the owner, Don, greets all new crews coming to stay at MDRS. He is also in charge of bringing water and other essential supplies out to the MDRS "hab." At Hollow Mountain, which as the name suggests is carved into a hollowed-out boulder, there was a display of books about Utah. Natalie purchased *Roadside Geology of Utah*, so that she could retroactively understand the geology we just drove through. This book focuses on the formations visible along the major highways in Utah. It confirmed that at mileposts 91 and 92 along Utah Route 24 the Morrison Formation hugged the road.

The next morning, before we set out for the field, Natalie introduced me to another geological representation. In her room (we slept in narrow chambers, about as long as a twin bed and twice as wide) she had pinned to the wall a large, beautifully colored topographic map. On the left were vivid reds and oranges. She pointed to where we were; a region colored in mossy yellows, greens, and blues. Each color corresponded to the dominant formation. The map was laden with other symbols, indexing who surveyed the area, what references existed in the geologic literature, where the fault lines were, and what period the formations were from. As a visualization it encompassed location, time, movement, and provenance. By locating us on the map Natalie was able to later narrate the time traveling we would be doing that day.

The map on her wall was only useful in identifying the top layer of rock. In the field, which was cut through with canyons, dry river valleys, and rifts, multiple formations were exposed, one atop another. On our third day in the field, we drove out to Angel Point, a lookout over one of the deepest canyons in the area. Stoker organized this excursion to find the Carmel Formation, which she thought might be of a suitable composition for testing the drill. We got out of the car at Angel Point and hiked a little way to the overlook. Several minutes of confusion followed during which Stoker, Natalie, Jon, Julia, and myself tried to figure out which layers were which. Natalie took out yet another visual tool, a stratigraphic column, which Buzz had given to us the previous day during our stop at the Bureau. It is a chart that displays strata in an ideal, vertical column, capturing the relative thickness of the different layers. The newest layer, the ground one walks on, is at the top. Along the left the layers are labeled with the corresponding geologic period. Most of the layers around MDRS were formed during the Jurassic period. As Rudwick points out, the columnar representation is a theoretical construct. Faults and folding are not depicted, and the strata are in an orderly horizontal position. This grammar of geology's visual language "is far removed from straightforward observation and . . . embodies complex visual conventions that have to be learned by practice" (Rudwick 1976, 166).

Even Natalie, fluent in this language, still needed a few moments to match the story on the stratigraphic column to the story playing out in front of her. Finally, she proclaimed, "OK, I know where we are." She had located us not in place but in time. Recognizing that we were standing on a formation from the Jurassic period brought the past into the present. She next trained us to see as she did, pointing to the lowest layer in the canyon and explaining that it was Wingate sandstone, the first and oldest of the Jurassic formations in this region. She built up the story from there; next came Kayenta, then Navajo; the thin uppermost layer, she pointed out, was Page, which meant that we were actually standing on the Carmel Formation. With a new understanding of our place, Stoker appraised the Carmel Formation. Even though there were several good drill sites, the location was too far away from the hab. It would be too long an excursion for a future crew in full simulation, dressed in space suits and unable to eat, use the bathroom, or scratch their noses.

The days spent with Natalie, who was scheduled to leave halfway through the first week, provided a lesson to all of us on how to see like a geologist. When I asked what I could do to hone my geological skills, she simply responded, "The more rocks you see, the better geologist you are. And that's just a fact." In a similar vein, Stoker remarked that when she is with Natalie she knows that she is not a geologist. Unlike Natalie, she cannot distinguish between the many similar-looking formations. However, even Natalie needed to brush up on her knowledge of this region. She had done field camps here as a student and so had a baseline familiarity. With the Roadside Geology text, she jogged her memory of the general lay of the land. The map on her wall helped fill in some finer details. Finally, it was the trip to the Bureau of Land Management, the personal guidance from Buzz, and the stratigraphic column that allowed Natalie to finally "know where we are." She learned to see the landscape and offered a geological narrative, which transformed what to us looked like jumbled or indistinguishable rock formations into a temporal and consequently spatial ordering.

Other visitors to MDRS report similar moments of making sense of the geologic landscape: "I sat on a mound of desiccated clay as the sun came up, both absorbing the light and contemplating the fact that the desert I sat in once was under an ocean. I tried to imagine sitting on a mound on the ocean floor. Imaginary fishes were swimming by, ogling me: the buttes and canyons, the red and gray slopes surrounding me were part of a far-off future" (Childress 2003). As the geological narrative juxtaposes the past and the present, one's imagination stretches to comprehend how this place that seems static now in fact has a dynamic past. The environment surrounding MDRS is thus more easily imagined as a place in the making.

The narratives Natalie uncovered at the quarry and from Angel Point remain just that, stories. Within geology a healthy amount of disagreement exists when it comes to interpreting formations. While it is true that every rock tells a story, for each person this story might be different. In an interview with planetary geologist Eric several months after this trip, he warned me about the subjectivity of geologic knowledge, especially as it pertains to geology on other planets. "I think everyone who wants to be a planetary geologist should have the experience of trying to map a small piece of the Earth," he suggested. On Earth, he explained, one can be standing on the surface with access to all the samples and equipment desired, yet there are often as many interpretations as there are people studying the geology of an area. However, with Mars, "people look at a few pictures and make proclamations that they understand how a place is based upon a very limited data set." If we draw from lessons of terrestrial geology, a claim of such definitiveness should be received with skepticism. The lesson we learn from Earth is not only that landscapes come to stand for multiple places throughout time but also that the narratives that organize a landscape's coming into being are multiple.

The first several days of our mission felt, to me, like a series of false starts. We sought out and found different formations, but to get to each was a journey of wrong directions, inaccurate maps, and being saved by the local guide, Buzz. Stoker was nonetheless content and optimistic about the reconnaissance. After all, she explained, these first days were all about "getting to know the place."

Landscape 2:
The Astronaut Geologist; An Astrogeological Narrative

The double exposure of Landscape 2 (fig. 1.4) juxtaposes a soon-to-be-enacted lunar scene against the stark backdrop of the American West. Astronaut Jim Irwin manipulates a scoop of soil, and Dave Scott looks on, as they prepare for the science objectives of *Apollo 15*, which was to launch in the summer of 1971. The canyon cutting through the middle ground

1.4 Jim Irwin (left) and Dave Scott (right) prepared for the *Apollo 15* mission with extensive training in field geology. The USGS mockup of the lunar rover is in the background. Image credit: NASA photo AP15-S71–23772; USGS Open-File Report 2005–1190, fig. 085b.

stands in for Hadley Rille, the lunar channel near which *Apollo 15* would soon land. Irwin and Scott were not hired for their geologic skills but, like almost every other astronaut, for their Right Stuff. Both had had distinguished careers in the Air Force and graduate-level training in aerospace engineering. How, then, did two test pilots find themselves in the Arizona desert, saddled with mock lunar equipment, identifying basalts? How did geology become the discipline for uncovering the story of the Moon and other planets? The emergent astrogeological narrative that makes sense of this multiple exposure of place is rooted in a midcentury debate over how best to know other places. Without the acceptance of this narrative, a place like MDRS would never have come to be. Also layered within this landscape is a second narrative: that of the frontier. The American frontier, though long settled, frames the training astronauts as they prepare to journey to the extraterrestrial frontier. Astronauts demonstrated that geologists and cowboys had roles to play on the lunar surface.

Historian Matthew Shindell (2010) importantly shows that geology's prominence in lunar science and later planetary science was not universally accepted. Rather it was contested within and without the discipline of geology. Shindell draws attention to the role the physical chemist Harold Urey played in questioning geology's place in solar system science. Urey viewed geology as a more qualitative field, insufficient (and unnecessary) compared to the quantitative results of his research. Further, Urey believed that the Moon was of a different origin from the Earth and the descriptive work of comparative geology would fall prey to misguided preconceptions (203).

Prior to the space program, the occasional geologist gingerly dipped a toe in the extraterrestrial waters. In 1938 the geologist Herman Fairchild, with apologies, suggested that geology might have something to offer the field of selenology (the study of the Moon). More daringly, in 1960, with breakthroughs in solar system science looming, Jack Green and Dael Wolfle suggested in an article published in *Science* that distinguishing selenology from geology set a dangerous precedent of coining a field and associated terms for each planet in the solar system. Instead, "*geology* and the *geo* terms can be extended from their earthly meaning to cover similar processes and features of other cosmic bodies.... Wherever they occur, a caldera is a caldera, sulfur is sulfur, and a reverse fault is a reverse fault" (1960, 1071). Geology, they argued, was not tied to Earth but was perhaps what tied Earth to other planets.

Also in 1960, Eugene Shoemaker became the first head of the newly established Astrogeology Branch of the United States Geologic Survey (USGS). Shoemaker passionately believed that geology had a great role to play in lunar and planetary geology. The most immediate task was to provide more detailed maps to aid in landing site selection, first for robots and later for humans. The geological community received with skepticism the remote reconnaissance he was doing on the Moon. Don Wilhelms, one member of Shoemaker's team, recalled giving a lecture at a French observatory in 1963 and explaining that you can tell the age of craters through visual inspection. The reaction to this claim was simple disbelief. Wilhelms retrospectively noted, "and that's been my experience through most of my career, especially in the sixties. It's become accepted now. There are a lot of people who now look at planets with this viewpoint. But in those days, it was like pulling teeth to get even a geologist to understand it."[10]

Several geologists, provoked by the application of geology to the Moon and a formal instantiation of this practice in geology's most prominent institution, the USGS, resisted the burgeoning astrogeology narrative and attempted to rein in geology as a uniquely Earthly discipline. Kalvero Rankama of the University of Helsinki protested: "I, for one, am taking strong exception to the use of 'geology' in [planetary geology, lunar geology, and astrogeology]. . . . Clearly, geology is restricted to the study of the Earth and of terrestrial phenomena and does not apply to extraterrestrial bodies and processes" (1962, 519).[11] The retort to this, which came a few years later, was that geology is not about Earth but earth. "It is the idea of the solidity, the stability of the land on which we can safely land after a travel through a more precarious medium, and it is the rock that composes this land." The analogy then stretched from early explorers who traveled through the seas to new land and astronaut explorers traveling through space to new, solid surfaces. "Geology will be the study of the place where they land" (Ronca 1965, 13).

After landing a man on the Moon, and after the astronauts received geologic training in large part due to the efforts of Shoemaker at the USGS, there was little hope for the dissenters to keep geology grounded on Earth. In a presidential address to the Geological Society of America in 1970, Morgan Davis began by declaring the past year "the most momentous year the geological profession has ever known. I refer, of course, to the lunar landing" (1970, 331). He used the opportunity to chastise those who sought to limit the scope of geology, urging his colleagues to embrace the study of the ocean floor, the Moon, and other planets. And by 1973 a retrospective of scientific work accomplished during the concluded Apollo program was careful to state: "This article attempts to evaluate the effect of the Apollo program on geology (using the term in its broadest sense)" (Smith and Steele 1973, 11). Geology, successfully but perhaps a bit uncomfortably, was now a science applicable to the Earth, the Moon, and in the near future, it was hoped, Mars.

The flourishing of the USGS's Astrogeology Branch helped legitimize geology as a multiplanetary science. In 1963, the branch moved from its original offices in Menlo Park, California (in the soon-to-be-established Silicon Valley), to Flagstaff, Arizona. They had better access there to telescopes and clearer seeing as they mapped the Moon. And, not insignificantly, their location was near Meteor Crater, where geologists had been doing fieldwork for some time to learn about impact geology as it might be applicable to the Moon. That same year, Shoemaker began training astronauts in lunar geology. He asked Wilhelms to be the primary lunar geologist and work alongside more traditional geologists in training the astronauts.

An astrogeologial narrative suggested that fieldwork, because of its grounding in geology, was a necessary skill. To teach fighter pilots how to identify rocks and make sense of their lunar surroundings, the USGS ran mini–field schools for the astronauts. After all, as one member of the Astrogeology Branch (who was not involved with astronaut training) put it, "fieldwork is the essence of geology, it really is. You know, you don't get any sense of the complexity of geology until you go out and try and do it."[12] This echoes Natalie's observation that the more rocks one sees, the better a geologist one is. Shoemaker and his colleagues had all received extensive fieldwork training in their geology educations and wished to impart this to the astronauts. Shoemaker wanted them to come out to Arizona, where the canyons and craters made ideal training grounds. He recalled his plan to "get scientist-astronauts away from their day to day involvement in the flight program and all the other activities the astronaut had, de-orbit them for awhile, get them on the ground, give them three months of really solid training—which is what it takes. You have to go out and do fieldwork to learn how to do fieldwork!"[13]

The aspiration behind astronaut training was not only to get them to see like geologists but also to teach them how to compare what they saw on

Earth to what they would see on the Moon. It was a move to place the terrestrial landscape on a different world. Donald Beattie writes, in his memoir *Taking Science to the Moon*, "on missions to the Moon some of the astronauts would comment on how much the Moon's surface looked like their memory of [their fieldwork sites]" (2001, 181). To understand the landscape, the astronauts overlaid Earth atop the lunar surface.

Partway through the Apollo program, after the initial landings were accomplished and as the future missions promised to be more science driven, USGS astrogeologists decided that the analog sites where they had been training astronauts were not sufficient. They decided to craft their own, ideal lunar landscape. After carefully studying a region of the Moon photographed by the lunar orbiter, they precisely placed explosives and produced just outside Flagstaff, to the extent possible, a facsimile of the lunar surface (Beattie 2001, 182–83). The Apollo training hugely influenced the ways the Moon and Earth were multiply exposed on each other's surfaces. The training activities supported an argument that geologic analogies work in both ways: the Moon can be understood through a terrestrial landscape, but also that a landscape on Earth can come to stand for the Moon. Further, being in the field at such analog sites is not only a legitimate but also a necessary way of knowing other planets.

Landscape 2 was captured on an Apollo fieldwork training expedition. It looks like a messy, multiple interpretation of a landscape. In a scene of shrubby desert with snow-capped mountains in the background, two men are encumbered by boxy backpacks with cameras strapped to their chests. They peer down at a mechanical scoop, inspecting a rock or soil sample that they might collect before returning to a minimalist buggy with a conspicuously large antenna attached. Looking at the picture now, we recognize familiar icons that tell us they are enacting the Moon on Earth. An astrogeology narrative that secures geology's utility in planetary science, which was by no means obvious or easily won, structures this landscape and brings dissonant images into harmony. It has also continued to be a powerful trope in contemporary planetary science and planetary geology, in which terrestrial fieldwork is understood as a proper tool for studying other planets.

Another motif that is unavoidable in space exploration rhetoric and central to Landscape 2 is the imagination of the American West and the frontier. Historian of technology David Nye (1997, 2003) describes a genre of

American narrative, what he calls stories of "second creation," to explain how people understand the way technologies shape landscapes and thus ways of living: "As Americans use machines to transform space into landscape (or to reconfigure an older landscape), they also construct narratives to make sense of this activity" (1997, 6). Nye further argues that the specter of the western frontier marks American narratives of technology, landscape, and place. The frontier, however falsely, conjures an image of emptiness that in turn justifies and even invites technological expansion. It is then no surprise that the astronauts who faced the "new frontier" of the Moon fancied themselves cowboys more than geologists. They were not explorers setting sail at the behest of the Crown; they imagined themselves setting out into the untamed wild of their own accord. A 1984 poster illustrates the enduring notion of the cowboy astronaut. It is a drawing of an astronaut floating in space, with the traditional astronaut suit supplemented with a cowboy hat, boots, and a branding iron. The caption of the poster reads "Texas — Still in the Frontier Business." Images like figure 1.5 circulated during the Apollo era. In this picture, a geologist (not an astronaut) tests out an early space suit. To keep the desert Sun off his head he dons a cowboy hat, creating an iconic image of the cowboy astronaut as embodied by a geologist. Training in the American West intensified the association between frontier and space, cowboy and astronaut.

The American scientist Vannevar Bush (1945) recast the frontier not as a physical entity but as something more mental in constitution when he called science the "endless frontier."[14] With the establishment of NASA and the goal of setting foot on the Moon, the cognitive and material frontiers were united.[15] Shoemaker, who found the frontier a compelling analogy for the growth of scientific research, recalled how this narrative ultimately unraveled: "I tried to draw — at the AAAS [American Association for the Advancement of Science] meeting — an analogy between the early exploration of the American West and the Apollo program. The difference, of course, was the early exploration of the American West led to evolving, continuing, growing scientific enterprise. The Apollo program didn't. But at that stage it wasn't clear it was going to happen that way. I was trying to say what we wanted to do was just build on this early stage exploration and go on to a really deep, meaningful program of scientific exploration."[16] Even while the Apollo program was still going on, Shoemaker began speaking out against NASA's vision of exploration. He rightly diagnosed NASA as an organiza-

1.5 The cowboy astronaut: geologist Joe O'Connor dons an early version of the Apollo spacesuit in the Hopi Buttes Volcanic Field, in the Territory of the Navajo Nation, Arizona, 1965. Image credit: USGS Open-File Report 2005–1190, fig. 031a.

tion of engineers, more interested in "can" than "why." After landing on the Moon, he did not think NASA should move quickly to Mars, the next target (Shoemaker 1969). Instead the Moon should be refashioned as a scientific frontier and explored in that nature. To prove that a destination could be reached was not, for Shoemaker, a satisfying ending to the story.

As Shoemaker seems to have predicted, NASA lost its role as primary story spinner for the frontier narrative. Grassroots organizations like the Mars Society and commercial ventures founded in the early 2000s picked up the thread. In the late 1980s, the members of the Mars Underground began planning missions to Mars without having articulated a thoughtful answer as to why one might want to do such a thing. After Robert Zubrin began attending the Case for Mars workshops, he formulated a justification for Mars missions that drew heavily on Frederick Turner's frontier thesis of the previous century. Zubrin's manifesto, "The Significance of the Martian Frontier," first appeared in 1994 in *Ad Astra*, a magazine of the National Space Society, and later in various volumes published following Case for Mars workshops (1994, 1996, 2000). Zubrin does not pull any punches when he writes: "The creation of a new frontier thus presents itself as America's and humanity's greatest social need. . . . Without a frontier to grow in, not only American society, but the entire global civilization based upon Western enlightenment values of humanism, reason, science and progress will die. I believe that humanity's new frontier can only be on Mars" (1994). Because Mars is remote yet able to be settled, Zubrin's logic goes, it is the ideal frontier. Zubrin's narrative of the New World as frontier suggests that in creating distance between the Old and the New, settlers escaped the aristocracy and could pursue their dreams of democracy. Humanity needs Mars because it will provide a stage for improvisation in order to manufacture the next great development. This narrative notably lacks natives and slavery, bigotry and disease, oppression and poverty. It is a powerful story because of its simplicity and because it cleanly juxtaposes alien Mars with the familiar frontier.

Geology, the frontier, the cowboy, the explorer, the Earth and the Moon all come together in Landscape 2 and are ordered by an astrogeological narrative. In this image, landscape and identity reinforce each other. The notion of the cowboy astronaut can make sense because of the western landscape. Doing geology on the Moon makes sense because astronauts train in the field on Earth. Geology might seem an odd addition to the "right stuff" arsenal, but geologists form their professional identities in landscapes not dissimilar to those where Apollo astronaut training occurred. "If I were to describe a classical excellent field geologist," a planetary geologist offered to me in an interview, "it would be someone who is extremely physically tough, and these are really explorer-style people. You know, people who can walk all day and don't get tired or whine when their backpack gets heavier and heavier because you're filling it with rocks. And at night you cook around a fire and sleep in a tent and even if the weather's bad you go out anyway." It is this formulation of the right stuff that informs fantasies of future Mars explorers.

Stoker and Jon told us participants at MDRS about the Association of Mars Explorers. There are about fifty members, and the founding document was written on a napkin at a dinner Stoker and Jon attended in 2002. According to the scrawl on this napkin, "it is a forum for explorers of the Martian frontier, including the deserts, mountains, and poles to inspire, astonish, and inform their fellow explorers with tales of courage, bravery, and ... discovery in the great frontiers of Mars or the Martian analog environments on the Earth."[17] They swap these stories every other year at a dinner held during the Astrobiology Science Conference. Jon teasingly described these dinners as a classic British gentlemen's club where they compare stories like "and then I saw a lion THIS big." Stoker came to the defense of the club, saying it is more like the Explorer's Club, which emphasizes scientific work. What is striking about this association, and its founding statement, is that members are recruited on the basis of both what they do and where they do it. The identities and landscapes central to this statement are part of a narrative that gained traction during Apollo training, is very much alive in the current activities at MDRS, and will only be resolved, so the story goes, when humans land on Mars.

Landscape 3:
Finding Mars on Earth; An Areological Narrative

On Mission Day 4, Natalie and Jon bid us and MDRS farewell. In the afternoon, Danny and Brian were to arrive, and Stoker, Devon, Julia, and I operated as a diminished crew that day. After an oatmeal breakfast to warm our bones following a cold night's sleep, the four of us locked up the hab (chaining the door closed) and spent the morning scouting more drill sites.

We were looking at different locations today, avoiding the rocky strata and examining clay and sand deposits instead. We turned off the main road, driving beyond "Historic Giles," a ghost town that appeared to be fashioned as a tourist site. Though there were no dwelling structures, there were a few mannequins dressed as outlaws and a wooden sign that welcomed us to Blue Valley Ranch and historic Old Giles Town of 1898. Another weathered post was engraved with a website to visit, but I discovered later that this website was also a ghost. Further down the road, another sign informed us that there were plots of land for sale. I looked out and saw desolation: a grey desert near a town that hosts a convenience store built into a boulder, a Chevron, and a burger shack. Stoker saw something else. Over lunch later that day, she voiced a desire to buy land here. I asked if she would build it up. No, she responded with a head shake. She'd just like to have it.

We pulled over at a possible drill site along the side of the road. There was a tall fortress of Mancos shale, a mesa, sitting atop a dune of sandy debris. Unlike the reds and browns of the Morrison and Summerville Formations, these features were grey and tan, more lunar than Martian. Though the fine sand made it difficult to maneuver, we collected some soil samples. Stoker and I implemented the collection and documentation procedure we had developed that morning. After about an hour at this site, I was heading back to put our equipment away in the truck when I heard sounds of excitement coming from Stoker and Julia behind me. I returned to see what they had discovered.

Scattered across the ground, a few meters from where we were collecting our samples, were hundreds of spherical rocks. Stoker and Julia eagerly collected samples, forgetting about the procedure we had worked on earlier and simply grabbing these rocks and putting them in whirl bags (fig. 1.6). Stoker looked satisfied and excited, exclaiming that exploration is fun. I tried to join in the excitement—what are these rocks? They are concretions, I was simply told. I responded enthusiastically, asking what that means. Are they made of something special or do they indicate something unique? The only response I got was a shrug, and Stoker and Julia went back to their frantic collecting.

After the initial wave of excitement died down and we were just standing around enjoying the site of the concretion field, I again asked why this find was so exhilarating. "These are like the blueberries on Mars," Stoker responded. Whereas the Apollo training taught astronauts to understand

1.6 Carol Stoker poses with a "concretion" at a potential drill site outside Hanksville, Utah. Photo by the author.

the Moon through the lens of Earth, an areological narrative was at work here that made sense of something on Earth by refracting it through knowledge of Mars.

Shortly after the Mars Exploration Rover *Opportunity* landed on Mars, it set its digital gaze on an outcrop of sedimentary rock near the rim of a crater. Sticking out of this rock were tiny globules, described by chief scientist Steve Squyres as resembling blueberries in a muffin. The description stuck, and the mystery of these blueberries occupied the science team for some time. Ultimately they concluded that these spheres were hematite-rich concretions, features formed when water carries dissolved minerals through softer rocks before settling within the rock. Eventually precipitates form in layers around the deposit, replacing the softer sediment with hard concretions. As wind erodes the softer material, the concretions end up littering the surface. Finding concretions on Mars was an exciting discovery indeed: the place *Opportunity* landed must once have been flowing with water (Squyres 2005).

Later in the day, we took a recreational hike through Little Wild Horse Canyon. As we walked along the base of the canyon, Julia spotted little spheres nested in the walls of the Navajo Sandstone formation. Stoker also identified these as concretions and noted that they are even more analogous to Mars blueberries. She referred to an issue of *Nature* published shortly after the blueberries were announced that mentioned these particular Utah concretions. This issue contains a letter and a separate discussion on this topic in the "News and Views" section. Using analogy between Mars and concretions found in Utah's Navajo Sandstone, the letter bolsters the link between hematite concretions and a watery past (Chan et al. 2004). The "News and Views" piece commenting on this letter, "On Earth, as It Is on Mars?," prominently juxtaposes terrestrial and Martian concretions, asking in the caption of the image shown in figure 1.7: "Earth or Mars?"

This case of concretions illustrates a complex trading between geological and areological narratives and how they ultimately complement each other to add clarity to an otherwise confusing landscape. When *Opportunity* first spied the blueberries, multiple explanations were offered based on terrestrial geology. After a battery of tests, the rover team reached consensus that they were concretions. Letters like the one in *Nature* used geological findings to support that these were concretions and the attending fluvial

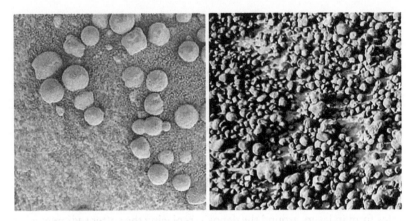

1.7 Martian (left) and terrestrial (right) concretions. Image credit: Catling 2004. Reprinted by permission of Macmillan Publishers Ltd: Nature, © 2004.

implication. Years later, with the connection between blueberries and concretions secure, Stoker was able to make sense of the Utah desert based on an areological narrative about concretions in the Martian desert.

Our crew was not the first to get excited by the discovery of Martian blueberries in the Utah desert. One crew from 2006 made it one of their scientific objectives to find blueberries while at MDRS. The crew found this task easier than expected and on the first extravehicular activity (as outings from the hab are known) radioed back, "Hab, we have found blueberries" (Gale 2005b). More than just setting crews on the hunt for blueberries, the landings of *Spirit* and *Opportunity* enriched the experience of being at MDRS by wedding the simulated activities to real and present events occurring on Mars's surface. The crew that was at MDRS during *Spirit*'s landing, as it began transmitting (and NASA began publicizing) daily photographs from the Martian surface, reported:

> As for the serenity outside the hab, it simply cannot be overstated. The amazing Mars-like terrain surrounding the hab has truly transported us, mind and spirit, to the Red Planet. Twice now we looked at pictures just received from Spirit and exclaimed, "Hey … that looks just like the area outside my window!" And the timing of this rotation, coinciding with the landing of Spirit, couldn't be better. As the world's attention gets focused on our dear neighbor, we're proud to be a part of the important work that will one day, hopefully very soon, enable scientists and

engineers, explorers just like us, to safely live and work on Mars. (Hegazy 2004)

The images from *Spirit* and *Opportunity* renewed the justification for MDRS and solidified the analog power of the place. On a subsequent crew one member reported a similar excitement uttered by one of her crewmates: "we saw outcroppings that looked exactly like the *Opportunity* and *Spirit* photos" (Wynn 2004a). Note that the analogy works both ways: the images from Mars look like the view outside the window, and the outcroppings look like images from Mars. The visual cues reinforce each other, leading to the doubly exposed landscape and the sensation of simultaneously being on Earth and on Mars.

Unlike the crews just mentioned, our crew unintentionally stumbled on the blueberries, offering a jarring moment when the terrestrial landscape was almost entirely replaced by the Martian, when one place became another. Consider a very similar account documented in a report a few years earlier. The commander of that crew (the first all-Austrian crew) wrote that their thirteenth day at MDRS was also his luckiest day. This is because "all was triggered by a special event in the afternoon, which switched my mind to Martian modus entirely and made me feel like a real astronaut on Mars." He had finally "land[ed] on the Red Planet, with both my mind and my heart." The moment occurred when, just as with Stoker and Julia, he was in the field. Also as with us, his mission was to collect soil samples. The previous two days had been dedicated to finding an easy path up to a point called Skyline Rim. From this point you can see the MDRS habitat to one side, and a canyon stretches out across the other. "It was at this precious moment," he writes, "that I suddenly really felt like an astronaut on Mars, when I remembered the graphic of the astronaut standing at the edge of Vallis [sic] Marineris with the rising sun in the back, and then suddenly the image switched and in the very next moment I was THIS astronaut" (Frischauf 2006). In being on this rim, the commander connected his embodied experience to a Martian feature, the deep and sprawling canyon Valles Marineris, and the landscape he was standing in ceased to be terrestrial. The painting he is referring to (and I will discuss more in the next section) is called *First Light* and was painted by Pat Rawlings, a prominent space artist, in 1988.

At the site near Giles, Stoker found a concretion nearly as big as her head. Even though the top had crumbled off, I asked if she wanted to pose with it. The image I captured (fig. 1.6) is a double exposure of Utah geology and Martian areology. There is a clear story that relates these concretions to concretions elsewhere. The areological narrative in some ways operates in a different direction from the astrogeological narrative. Both suggest that landscapes in one place are tied to landscapes in another place. However, whereas the astrogeological narrative uses Earth to make sense of the Moon and other planets, Stoker employed an areological narrative to make sense of Earth based on her knowledge of Mars. Finding concretions on Earth was an exciting discovery for my crew not because it confirmed that water had once flowed here (there was a stream not far away) but because it was a symbol of Mars. It offered Stoker, who knows far more about Martian geology than terrestrial geology, a new connection to the landscape. Though the conceit at analog sites such as MDRS is that we are studying the Earth in order to better understand another place, in actuality when an areological narrative was available, we appreciated these terrestrial features because of our knowledge about Mars. The concretion field became a place-world, as Basso might say, not by invoking past geological or human events but by invoking knowledge of the present, albeit the present of an entirely different place separated by the extremes of space.

It is not uncommon for planetary geologists to undertake terrestrial fieldwork similar to the work we were engaged with at MDRS. At NASA Ames I spoke with several planetary geologists for whom fieldwork informed their research. When I sat down with Eric, one of the first things we discussed was his training in geology. I asked about his experience of field camp during his education, and he made it clear that fieldwork was not just part of his training but part of his current practice. To prove this, he stated that he just got back from the field two weeks ago. As he said this, he dropped a black spiral notebook that was sitting on his desk onto the table between us with a hearty thump. This, I took it, was his field notebook—a ubiquitous tool of the geologist and anthropologist alike.

He was working on a project about meandering rivers. On Earth, a meandering river forms where the river walls are reinforced with vegetation. On Mars scientists have identified meandering rivers in satellite images. These "rivers" are not flowing with water today, but the pattern on the dry surface suggests this was the case in the past. Though it would be nice for astrobiologists to announce Martian meanders as proof of vegetation, Eric and his collaborators want to investigate alternative mechanisms for meanders. To think through this problem, they searched for an appropriate Earth analog—a river in a more desiccated landscape with scarce vegetation. This brought him to the Quinn River in the Black Rock Desert of Nevada. I wondered if while in the field he and his team ever imagined they were on Mars; did they ever think that they were directly studying Mars? "Sure," he responded. "I mean it was a topic of discussion." But they were always conscious of teasing out what in the landscape was purely terrestrial and what might be applicable elsewhere.

Eric, who finds the merging of the Martian and terrestrial landscapes to be a productive tool in his research, thinks that planetary science is at a disadvantage because of the unwillingness by most people to go into the field "and see geology as it is." Another NASA planetary geologist, Eliza, corroborates this view. She works on dune processes on Mars. When I spoke with her, she was in the process of organizing a workshop on planetary dunes. She was keen to organize a little field trip as part of the agenda, just to get the planetary people to see dunes "in real life." To "go out there, walk around the dunes for a couple hours. It doesn't take much, just to get some idea. OK, that's how tall they are, this is what little ripples look like, OK. Yes, that stuff actually happens." This kind of exposure teaches one to see dunes in a different way. She also mentioned that she was taking her summer students to the Mojave for an orienting dune trip, "just to show them. There's nothing like seeing them and walking around on them. After you know a little bit about them. . . . Suddenly you might go and walk on a dune and say oh, this is fun. But when you actually know what to look for, suddenly there's a lot more detail there than you ever thought."

Planetary scientists go on these fieldwork trips in an attempt to locate Mars in a place they can actually travel to. Fieldwork for geology and anthropology is grounded in a notion that "being there" is a valuable and telling experience. I will return to this notion of being there and its resonance with anthropological methods in chapter 4 where I discuss astronomical trips to observatories. What distinguishes the fieldwork I discuss in this chapter is that for this "analog" work in planetary geology, being there is complicated by the fact that the "there" the scientists travel to on Earth is not the "there" they are actually seeking to understand. For being in the field to be epistemologically fruitful, scientists must persuasively argue that there is a *there* there that evokes Mars. My own experiences in the field sometimes lagged behind those of the people I was with. For example, Stoker saw the concretion field as a "there" similar to Mars well before I did. While I was witnessing scientists at work in the Utah desert, Stoker and Julia were already infusing that place with Mars. Our fields momentarily diverged.

Stoker and other planetary geologists enter the field with more knowledge of areology than geology and consequently understand the terrestrial structures through the lens of Mars. At the same time, they expect that being at these structures will offer a new intimacy with the distant planet, an intimacy that "being there" affords. They are constantly using dunes, rivers, and concretions to draw links between distant worlds. It is this focus on specific features that transforms Mars from a planet into a landscape. Eric narrated his changing understanding of Mars. "You can't be much younger than I am," he began, "and remember what the solar system was imagined to be before spacecrafts starting going there." He lived through the release of the first photographs of Mars in 1965. But it was not until he saw images of the surface and was able to identify craters and plains and volcanoes that he began to relate to Mars. "It literally changed my psychological disposition towards them. Before seeing them, before they became landscapes, they were astronomical objects. I could go out and look at them with my telescope. And now they are geologic landscapes that I look at with spacecraft images." For this geologist a planet became a place when the alien object resolved into a terrestrial-seeming landscape. When I asked him if the gas giant planets had undergone a similar transformation and become places, he was clear to specify that while they are places, "they are not geological places" because without a surface there is no way to do geology on them. While Jupiter is pretty, and he can appreciate it when looking through a telescope, he categorizes it more with the Sun than with the rocky planets and satellites he studies. He returned again to Jupiter when describing why he found planetary geology interesting. Jupiter's "timelessness makes it inherently less appealing, then, [and harder] to be able to tell a story about the place." Without a temporally shifting landscape, Jupiter lacks a narrative.

Geologists are piecing together a narrative about Mars based on the shapes of dried-up rivers, the tumbling of craters, and the presence of anomalies like concretions. They cannot physically place themselves in the Martian landscape and thus seek out terrestrial landscapes to inhabit. This practice allows the planetary geologist to investigate the story she is crafting and in so doing gives new meaning to the landscape at hand. The areological narrative draws attention to features that the geological narrative might miss.

As Julia, Devon, and I piled back into the car and prepared to leave the concretion field, Stoker took in this particular Utah landscape one more time. She sighed before exclaiming, "Really Beautiful. What *a place!*"

Landscape 4:
Living and Working on Mars; a Science Fiction Narrative

The landscapes so far discussed were images from the field. They captured double exposures of a single location over time (as with the geological narrative) and the blending of places removed in space but existing in the same moment (the astrogeological and areological narratives). Our days were mostly spent in the field, navigating these multiple places. In the morning and evenings we dwelt at the hab, and there a different narrative ordered our practice. I call this the science fiction narrative. Utah and Mars come together at the hab in a different way than they do in the field. The Mars imagined within the hab and nearby is not the scientifically known Mars of the present but the human-inhabited Mars of the future. It is a vision of Mars shaped by popularizers and science fiction writers who have speculated on the nuances of human life in a Martian colony. The hab becomes a place-world not because of remembrances of the past but through invocations of a "future past," a phrase used by literary scholar Istvan Csicsery-Ronay to name an aesthetic of science fiction (2008, 76). We can think of sci-fi stories not as prophecies but as histories of events yet to happen. Those familiar with the sci-fi genre, as most MDRS visitors are, have the opportunity to weave their personal narrative into a familiar yet speculative story.

Narratives of science fiction spill over into the library and DVD collection amassed at the hab over the years. Most crews reserve the evening for entertainment: watching movies on laptops or, as my crew did, bringing a projector for a more theatrical experience. Crews report mostly watching science fiction, ranging from classics like *2001* and *Dune* to the more recent campy fare of *Starship Troopers* and *Galaxy Quest*. There is a healthy amount of Mars-based films, like *Total Recall*, *Mars Attacks*, and *Red Planet*. Shortly after the one and only season of *Firefly*—a television show that fused western

1.8 Carol Stoker, Julia, and Brian, wearing simulated space suits, walk back to the MDRS habitat after an afternoon of field tests. Photo by the author.

frontier and sci-fi tropes—was released on DVD, crews frequently reported evening viewing of episodes (my crew included).

Just as knowledge of Mars helped Stoker make sense of the landscape, knowledge of science fiction serves for certain crew members as an explanatory device. One visitor to MDRS in its first season found himself the first to arrive and compared the experience of waking up alone in the hab (which he referred to as a spaceship) as reminiscent of a *Twilight Zone* episode (Burbank 2002). Another crew member, a year later, described the confluence between the sci-fi novel he was listening to (*Mars*, by Ben Bova, about the first six-person mission to Mars) and his experience at MDRS. As he was driving back to the hab from an emergency trip into town to get spare parts for the generator, he wrote, "the wind had really picked up, and there was blowing sand across many sections of the road. Just by chance, the characters in 'Mars' were experiencing high (200 kph) winds and zero visibility while on an excursion in their rover. Made me feel right at home" (Fuller 2003). In other crew reports references to *Star Trek*, *2001*, and *The X-Files* were used to make sense of the day's events.

As when the areological narrative interceded in the concretion field, the science fiction narrative sometimes causes a similar shift in how the landscape is perceived and its accompanying sense of place:

> I did the dishes this morning and would like to thank the person who had the foresight to put a porthole window right over the kitchen sink. What a view, you really picture yourself looking right at the red planet while wiping your plates and pots. In fact, I was day-dreaming a bit the other day. Jim Barba had suggested that we might need a new diode for Casper [the generator]. I was pondering this while doing dishes, when suddenly everything shifted just slightly and I thought I must quickly obtain a dilithium crystal [a reference to a *Star Trek* technology], but first I should hang the pots out to scour in the wind. Then an image of Harcourt Fenton Mudd [a criminal in *Star Trek*] crossed my mind and I quickly flashed back to reality. (M. Zubrin 2006)[18]

Looking out the same porthole window while making pancakes from Bisquick and powdered milk, Danny, a member of my crew, one morning exclaimed, "Ah Mars. I love looking out at Mars." Danny was representative of a typical inhabitant of MDRS, being a space exploration enthusiast and possessing a deep knowledge of science fiction and other "nerd" cultures.

He was tall and lanky, with long brown hair often hidden under a bandana. He mostly wore black jeans and T-shirts, a mark of his involvement with technical theater during college. While at MDRS, Danny had to attend a virtual meeting of Students for the Exploration and Development of Space, which has chapters at approximately thirty universities; he was on the executive board for the U.S. branch of the organization. He had just finished his undergraduate degree in aerospace engineering, with minors in astronomy, physics, and planetary science. Before starting a master's program in space studies he was interning with Stoker for a semester. Danny had also been to MDRS several times before, both as part of a different, NASA-affiliated crew and as part of the engineering crew that polishes up the hab at the beginning of the season. He felt a great deal of ownership over the hab and spent much of his time at MDRS doing repair work—fixing the telescope in the Musk Observatory[19] and maintaining the green water system that allowed us the luxury of a flush toilet.

In a crew with a healthy appreciation of science fiction, Danny's knowledge of the genre outshone us all. He read Robert Heinlein during breaks and had heated discussions with Stoker over whether *Battlestar Galactica* or *Babylon 5* was the better show. His prowess in the genre spurred Stoker to call him out as a true geek on more than one occasion. It was therefore Danny to whom I went in order to decode the science fiction references of the hab. I asked about the meaning behind the flag that waved from the roof, a sequence of red, green, and blue rectangles (visible in fig. 1.8). Danny told me that it is a homage to K. Stanley Robinson's trilogy *Red Mars, Green Mars, Blue Mars* and a general reflection of the ethos behind the goal of colonizing Mars. Red Mars is the Mars of today, inhospitable to human habitation. Green Mars suggests a time after colonists have "terraformed" Mars, when vegetation is sustainable, the climate warmed, and the atmosphere thickened. Blue Mars is a habitable Mars, with a water cycle that both keeps rivers flowing and rain pouring.

Robinson's epic trilogy, in particular the beginning of *Red Mars*, reads at times like a description of the MDRS facilities. The saga begins with the first one hundred colonists journeying from Earth to Mars. Arkady, the ship's anarchist, insists halfway through the journey that they scrap the current architectural plans for the colony and remake one based on equality and the fusing of work and life. He puts forth his vision for living on Mars: the habi-

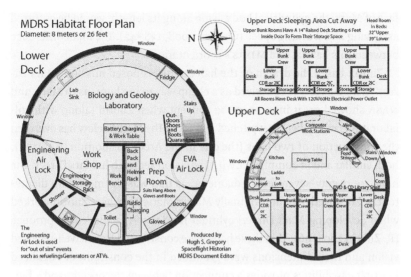

1.9 The MDRS habitat layout. The equal-sized rooms are set in the rim, with common space making up the rest of the second floor. The first floor is devoted to engineering and science equipment. Image credit: Mars Society.

tat should be circular or a geodesic dome, and "as for the insides, perhaps mostly open. Everyone should have their rooms, sure, but these should be small. Set in the rim perhaps, and facing larger communal spaces. . . . This is architectural grammar that would say 'All equal'" (Robinson 1993, 80). Sure enough, MDRS is a cylindrical construction; the first floor is designated as the workspace, with engineering and biology facilities. The second floor has rooms of equal size along the rim, opening up into the communal kitchen/living room/den/study (see fig. 1.9). The hab is not an alien thing of the future but a recognizable living space for many crew members: "The location of the hab is perfect. We might as well be on Mars! The sight of the hab would be familiar to anyone who has read books on Martian exploration by humans, and yet it seems so much smaller than I had imagined" (Verma 2006).

The writings of Robinson and the imagination of this crew member were likely influenced by early feasibility studies going back to the Case for Mars conferences. During the 1984 one the attendees broke into small teams and considered mission and habitat architectures. The lander designed there was cylindrical so as to fit within the nose of a nuclear warhead. On the surface of Mars, the cylindrical body, lying along its length, would be outfitted as a living and working space (Chaikin 2008, 138–42). The main difference between this habitat and MDRS's is orientation.

Zubrin based the design of the hab on his proposed mission, Mars Direct. Mars Direct was conceived as a cheap way to get humans to Mars, and NASA showed great interest in the concept when Zubrin (along with collaborator David Baker) first pitched it in 1990. The Mars Society has overseen the construction of two habs (the one in the Arctic in addition to MDRS), both similar and in compliance with the vision set forth in Mars Direct. The habs are twenty-seven feet in diameter, as that was the maximum diameter the Lockheed Martin (originally Martin Marietta, where Zubrin worked when developing Mars Direct) cylinder fabrication facility could produce (R. Zubrin 2004, 96). That these habs were constructed from a well-received vision and their dimensions were grounded in the constraint of Lockheed Martin's capabilities provides a connection between the present and a fantastical future. This present, however, is slowly moving in a different direction. If humans actually do travel to Mars, it will not necessarily be on ships manufactured by Lockheed Martin but perhaps instead on platforms engineered by the growing commercial space flight industry. Regardless, that the hab is constructed from blueprints of the present aerospace industry supports the sci-fi trope of future history suggested by Csicsery-Ronay in which the present becomes a prehistory of the future.

The space suits that we occasionally donned were also directly out of Robinson's *Red Mars*. We put on the "fairly standard hard helmet, and locked it to the suit's neck ring; then shouldered into an airtank backpack, and linked its air tubes to [the] helmet" (Robinson 1993, 128). Our helmets were fashioned out of garbage can lids and plexiglas; our backpacks contained a standard fan, which did little to infuse the stuffy helmet with fresh air. And though putting on the suit was a novelty and made us laugh, it rang familiar.

The image of Landscape 4 in figure 1.8 was captured toward the end of the mission. Most of the collection and testing tasks had been accomplished, and Stoker suggested it would be a good idea to test out the EVA (extravehicular activity) suits. Julia and Brian gleefully volunteered to suit up along with Stoker, and the three costumed explorers decided to test the dexterity of the suits by doing some soil sampling in situ (fig. 1.10). In Landscape 4 one can imagine an isolated habitat and the relief that returning

1.10 Simulating working on Mars in Utah. Carol Stoker, Julia, and Brian crouch around an instrument that tests the mineral properties of soil. Photo by the author.

to it would bring. One could finally push the hair out of one's eyes, use the bathroom, and breathe deeply. The blue sky and an airplane's gaseous trail toward the top of the frame remind the viewer that this is not Mars.

The first crews of the arctic Mars simulation emphasized the double exposure of place by using an orange or red filter when taking photographs to suggest to the viewer that they are actually looking at a peopled Martian landscape. In a spectacular convergence of the imagined future and present experience, NASA's Pascal Lee (Zubrin's original partner in establishing the arctic Mars station) enacted Pat Rawlings's 1988 painting *First Light*. It depicts two astronauts, one gazing across a canyon and another repelling down the side of a Martian cliff. After finding a suitable cliff near the arctic Mars station, Lee orchestrated a red-tinged photograph in which he repelled down the cliff in a simulated Mars Society space suit. As science writer William Fox observes, the Rawlings painting "was in itself an imagined re-creation of an event that has yet to happen in the future" (2006, 160). Thus, Lee's photograph is a present depiction of a future of the past. It is a photograph of a landscape waiting to be developed.

The science fiction narrative organizes these connections across time and space. It provides a framework already familiar to those at MDRS. For those saturated with the tales of Arthur C. Clarke and Ray Bradbury, Philip K. Dick and William S. Burroughs, living for a couple of weeks in the hab is far from out of the ordinary; it is familiar. Just as the Utah desert made the most sense to Mars scientists once elements of Martian geology were present in the landscape, this cylindrical living space makes complete sense when viewed through the lens of science fiction. For those who have spent decades reading about future colonies on Mars, it is a joy to bring those elements into the present. Numerous participants write how living at MDRS for two weeks allows them to enact long-imagined narratives: "Yep, we admit it. On our first day here, we took photos of everything, including each other, especially when we made our first EVA (extra-vehicular activity) today, wearing our space suits for the first time. But who could blame us? Not many people have been to Mars yet. For some of us, coming to MDRS fulfills a lifelong dream, or at least the closest we will get to it in this lifetime. We could hardly stop ooh-ing and aah-ing over the fantastic rock formations, the multi-colored hills, the beautiful crisp sky, and the stillness" (Wynn 2004b). It would seem that the hab and its sci-fi heritage were more familiar environs than the alien Utah desert.

Conclusion:
A Utopian Narrative?

The landscapes of MDRS are informed by multiple ideas of temporality and spatiality. A geological narrative brings former iterations of the place into being; an areological narrative, built atop an astrogeological narrative, telegraphs the present of another already scientifically investigated world; and a science fiction narrative imparts a vision of the future. If there is a single narrative that orders these juxtapositions of place and time, it is perhaps one of a scientifically driven, harmonious occupation of another planet. This is the foundation of perhaps a utopian story. How then does a utopian narrative order the imaginations and materiality of MDRS?

Karl Mannheim (1929) suggested a binary relationship between ideology and utopia. Whereas an ideology maintains the status quo, utopia—as a system of ideas—works to change the status quo. Literary scholar Tom Moylan points out that this oversimplifies the relationship between ideology and utopia. For Moylan, utopia operates within an ideology, both supporting and challenging it (1986, 19). This was quite literally so in the case of the MDRS crew I joined. The mission of MDRS is to pave the way for the human colonization of Mars. This is a utopian narrative borne out of a frustration with NASA's space exploration ideology, which has long stalled a human mission to Mars. Yet the crew led by Stoker, populated mostly by NASA employees, represented an effort to make room for the utopian fantasy of MDRS within the NASA regime.

It is possible that the contradiction between MDRS and NASA missions caused no problem in the minds of my crew because MDRS resonates with a utopia, a "no place" both in time and space. Being at MDRS offers an artifice of isolation, as crew members are removed from their work and home and are without many of the trappings that structure their understandings of contemporary life. Each crew decides to what extent they want to enhance the feeling of MDRS being a no place. Cell phone reception was spotty, so most of us volunteered to leave our phones off for the mission to simulate a remove between MDRS and elsewhere. Though we had Internet access, the bandwidth was limited and forced us to adopt different usage habits. Crews are told that water is scarce and to shower conservatively at the hab. Our crew adopted an every third day shower routine, but I found out toward the end that one crew member had been showering every day. Danny on the other hand sought a more "authentic" experience and opted to shower only once. We were simulating isolation, but was this Mars outpost really a utopia?

Robert Markley explores the complex vision of utopia that Robinson puts forward in his Mars trilogy.[20] At a broad level, Robinson classifies science fiction as "historical simulations" (quoted in Markley 2005, 355), meaning that science fiction does not *represent* experience but *simulates* the possible future of experience. Simulation does not imitate; it generates. Markley uses this distinction between representation and simulation, formulated by Steve Shaviro, to suggest that the "'utopian' possibilities of science fiction occupy a register of simulation: they give imaginative form to the desire to think beyond the contradictions of historical existence, and … beyond our location in time, culture, and geography" (356). As a simulation, MDRS does just this. It is a generative medium on which to speculate about future Martian experience. It is a material instantiation of a speculative future. It creates a history even as it simulates the future.[21]

The landscapes I have structured this chapter around certainly draw connections to different "times, cultures, and geographies," but these connections are not exclusively beyond the present. Instead of a utopian simulation, a no place entirely other, MDRS is very much of the current time, making connections to elsewheres and other times. Perhaps it is better thought of as a Foucauldian heterotopia (Foucault 1967). The successful utopia is one that is entirely separate—surrounded by a moat or part of an undiscovered land. A heterotopia, which Foucault contrasts to the imagined utopia, is somewhere that appears as no place, but is actually every place. Foucault's example par excellence is a boat: while it seems bounded and isolated, in practice it extends over the whole ocean and connects every port and every facet of human presence. For MDRS, these connections are made through a series of narratives; MDRS is heterotropian. The utopian narrative might appear as a master narrative, but it contains stories of geologic history, the ideal of fieldwork, the frontier and the American West, and scientific and speculative stories.

Whether MDRS is imagined as a utopia, a heterotopia, or a heterotropia, living in such close quarters ultimately becomes dystopian—a realization of the different expectations at work in one place. One of my crew's most heated discussions happened after dinner one evening. We were sitting in the common space around a folding table surrounded by the remnants of missions past, which leave their mark in the form of mission badges, DVDs, books, posters, Post-its, and makeshift signs conveying tips on living in the hab. We started brainstorming ways to improve that living. I observed that the clutter of the main area gave the hab a sense of hominess that welcomed modification—and these modifications were rarely improvements. Devon further suggested that to avoid users constantly altering the hab, the living space should be as sterile as possible. Danny, who has a deep affection for the quirkiness of the hab, leaped to defend the personality of the interior. Besides, he said, people will continue to behave as they have. We disagreed, suggesting that if the environment is clean and if directions are clear, people will comply.

This disagreement over the interior of the hab spoke to the different purposes we saw MDRS playing. Danny saw it as a retreat, a place to tinker with scientific goodies and imagine another way of life. That the hab was somewhat "hacked" together was not at odds with his vision. Devon imagined the hab as a machine with which to test equipment; the more standard and streamlined the better. For Stoker, MDRS is animated by both of these views. Having a place to fool around with the imagination of what it is like to be on Mars was a welcome departure from NASA work, but MDRS also needed to be a place that was producing science, and for this to be achieved a smooth organization behind the hab was needed.

MDRS is multiple, in landscapes, stories, temporalities, and purposes. But if there is one thing that is singular across experience, it is that it allows one to connect with Mars. Even I, with no desire to travel to Mars or even a strong commitment to human space exploration, found myself wondering just how faithful this simulation was; if this was really how people would live on another planet. I imagined myself on Mars. Whether such an imagined experience is fleeting or is sustained for the better part of two weeks, MDRS enables a double exposure of Mars and Earth. It allows for an abstract and distant planet to become something traversable and capable of up-close study. The many narratives woven through words and daily actions make present and tangible a planetary place and ground the planetary imagination in a small swath of the Utah desert.

I began this chapter with scenes of arrival and will conclude with a tale of leaving MDRS. This narrative comes from Devjit, an undergraduate I worked with while at MIT (see chapter 3). He went to MDRS during the season after my mission with a crew assembled from his summer experience at the NASA Academy. He did not know his crew very well beforehand but told me how well they got along at the hab. They were all eager and excited to carry out a high-fidelity simulation. Devjit told me that this crew decided to use the hab's Internet connection but to disable all chat clients, thus eliminating real-time conversation with anyone not at MDRS. A few days in they lifted their no Facebook ban but silenced its chat feature. They turned off their cell phones and allowed themselves to create a little community for two weeks, cooking, cleaning, and working together. It can be a shock to leave the hab at the end of the mission, to abandon the space suit and exit onto Utah soil as opposed to Martian terrain. Devjit, like astronauts returning from space, reflected on the increase of "Gs" (in his case cellular, not gravitational) on reentry: driving away from Hanksville, Devjit turned his cell phone back on and watched as it accumulated bars and upgraded from 0 to 1 to 3G. Devjit described an awareness of gradually becoming more connected. This was bittersweet, as with greater connection came greater distance between crew members. The story created by his community over the past two weeks was unraveling. People were being pulled back into their "real" lives.

The power of MDRS is that it brings together elements of different times and different places, enabling its occupants to weave together any number of narratives. Capturing snapshots of the landscape—double and multiple exposures of the surroundings—momentarily holds steady the shifting social, scientific, historical, and literary processes by which Mars is made as a place on Earth. Even as the landscapes change, elements of the four narratives I heard at MDRS are retold after scientists return from the field to their laboratories. In the next chapter I show Mars being made as a place using a different set of techniques—mappings; but narrative threads, familiar now to the reader, will continue to give form to a planetary imagination.

M A P P I N G   M A R S

I N   S I L I C O N   V A L L E Y

On August 6, 2012, the rover *Curiosity* landed on Mars. The news media exclaimed over the detailed pictures returned from the alien surface. A panorama of Mount Sharp, the rover's destination, portrayed a boulder-ridden, mountainous landscape that was likened to the Grand Canyon. These spectacular images that enlivened the public Martian imagination were business as usual for Mars scientists. Since the early 2000s, scientists have been studying similarly jaw-dropping images day after day. All of these photographs are available for the public to download, yet it is only during a high-profile event, like the landing of *Curiosity*, that nonscientists chatter about and fawn over such images. A few years before this landing I had sat in on a meeting where software developers brainstormed how to circulate more widely the detailed photographs already returned from the 2003 rovers and high-resolution Mars imaging orbiters. How could they get nonscientists to see Mars in the way scientists do, as an exciting and changing place? To capture how Mars is made as such requires not only examining how images of the planet are presented and circulated but also understanding the technical work needed to produce Mars as a world.

To attend this meeting, I had flown from San Francisco to Seattle with Janice, a business development manager at NASA Ames, who had pulled some strings so that I could attend this meeting, which was held at Microsoft Research, the computer company's R&D branch. The meeting was a chance to touch base on a collaboration in which NASA would deliver to Microsoft the highest resolution, interactive map of Mars ever produced. A small group at NASA Ames, known as the Mapmakers, was in charge of building this map for Microsoft's web and desktop application WorldWide Telescope (WWT). I was a participant observer with the Mapmakers, and had already been at NASA Ames for a month working on producing public outreach tours of the Mars map in WWT.[1] The meeting was attended by, in addition to me and Janice, the chief information officers of NASA Ames and the NASA Goddard Space Flight Center, marketing specialists from both NASA and Microsoft, and of course the engineers in charge of building the software. The engineers from Microsoft were responsible for the whole WWT system. From NASA, the Mapmakers who attended the meeting were three computer scientists in their late twenties and early thirties.

The night before this meeting I met up with Janice and the Mapmakers — Aaron, Seth, and Jesse — in the lobby of our boutique hotel. We sat in a mechanical wilderness: the columns of the room were made up of tree trunks spliced with iron rods, limbs attached with metal joints. Seth jokingly described the aesthetic as "technorustic." This seemed a fitting ambience as we sat in a circle, laptops perched on laps or coffee tables, and the engineers digitally tamed the Martian landscape in preparation for tomorrow's demonstration of their map. At the start of the work session Jesse and Seth tried to figure out a bug in the code, while Janice and Aaron, the director of the Mapmakers, hashed out a business plan to present the next day. After the conversation Janice said good night, and Aaron joined Jesse and Seth. At around 1 A.M. I excused myself as well, as the engineers had settled into a silent, nearly meditative coding session that was sure to last several more hours.

At Microsoft Research the next day, Aaron (who showed no signs of the previous late night) led the demo. His technical task was to explain the data processing needs necessary to stitch together and "serve up" the terabytes of NASA Mars imagery as a seamless globe. However, to excite the room, Aaron began with a tour of the still-in-progress map. Projected in front of the room was an image of a Martian crater. The image was so detailed that Aaron was able to guide our eyes to a dry riverbed cutting through the crater wall. As he panned around the image, he explained, "I chose this [region of Mars] because it's really interesting to explore." He zoomed out from the crater so we could see more of the Martian globe. With the press of a button, Aaron turned on the terrain model, which instantly rendered Mars in 3-D. He navigated the map such that it felt like we were in orbit. The ridges of craters and peaks of mountains seemed to rise up from the surface. Aaron flew us toward the North Pole and again zoomed so far into the surface that we were able to make out individual boulders. At maximum zoom we discerned two dark smudges. With a smile Aaron explained that one smudge was the Phoenix lander, from the mission that landed in 2007 to assess the habitability of the Martian arctic. The other smudge was the lander's discarded heat shield. He was showing us Mars on a human scale. In Aaron's hands Mars was a place very much alive and filled with mystery. The general public would soon be able to experience this vision of Mars in 3-D on their home computers.

This chapter tracks how the planetary imagination that guided Aaron's presentation and that I showed being enacted in chapter 1, an imagination in which Mars is positioned as a site of exploration, becomes mobile and shared beyond the scientific community. Specifically, I focus on the digital globes of Mars that the Mapmakers produced for Microsoft's WWT and Google Earth. These free web and desktop applications, marketed to the lay user, offer an interactive and immersive experience in which the user virtually navigates the Martian (and terrestrial) globe, zooming in, panning across the surface, and tilting his or her perspective. These programs, which facilitate the circulation of today's Martian planetary imagination, seem new—they are high-resolution and immersive global representations—but are in fact reworkings of traditional mapping technologies that have long been used as tools for cultivating a planetary imagination of Earth.[2] Maps, even scientific maps, are not neutral representations but are infused with the norms of the community that creates them (Harley 1989; Wood and Fels 1992; Edney 1997; Pickles 2004 ; Crampton 2009). The places that maps create are not predetermined but are an outcome of cartographic work. Critical cartographers have thus examined how mapmaking and map using are processes, divesting the map of any fixity. Maps are always becoming (Del Casino and Hanna 2006; Kitchin, Gleeson, and Dodge 2013).

And if maps are always becoming, so are the places they aim to represent. The Mars depicted in today's maps is different from the Mars of last century's maps. In the late nineteenth century astronomers drew maps in specific ways to argue for an inhabited Mars. Richard Proctor's 1869 map of Mars labeled features with terrestrial terms (ocean, continent) to mark it as similar to Earth and thus possibly filled with life. Further, Proctor chose a Mercator projection (a standard terrestrial projection) to emphasize that Mars was not an object but a place that was indeed familiarly mappable and navigable (Lane 2010, 27). More famously, Percival Lowell's maps contained straight lines connected by nodes to bolster his theory that Mars had canals and thus was inhabited by life forms intelligent enough to build a complex irrigation system. The primary goal of today's maps is not to emphasize the potential for alien habitation but to establish Mars as inviting to human explorers.

To achieve this goal, the work of the Mapmakers necessarily strives to offer a coherent, indeed robust, image of Mars. In particular, they present Mars as (1) *democratic*, meaning accessible to all, (2) a place of experience, exemplified by the *three-dimensional* map, and (3) *dynamic* and therefore worthy of continued scientific study. These aspects establish Mars as both a place and, more important, a destination. They stabilize Mars and thus facilitate the propagation of a planetary imagination. But this relationship between mappings and place-making is not without contradictions. Already one can see how even as the Mapmakers promote the dynamism of Mars, they do so in the service of fixing it in the user's mind as a certain kind of (exciting, explorable) place. Similarly, the democratic ethos and desire for openness (both of mapping tools and data, as I shall describe) exist alongside a state project that constrains how exploration occurs. Finally, the 3-D technology that facilitates an immersive experience invites a sense of the *real* even as Mars remains emblematic of what Baudrillard (1983) has called the hyperreal, in that the map precedes the territory.

These contradictions emphasize that mapping Mars is not as straightforward as it might seem. To even assert that Mars can be mapped in a way similar to Earth is already the beginning of an argument about what kind of place Mars is. The mapmaking project I discuss in this chapter naturalizes and makes widely available a planetary imagination that supports the resilient goal held by NASA and space enthusiasts to eventually set human foot on the awaiting Red Planet.[3] This project, though targeted at a distant world, is very much a product of its terrestrial environment. The democratic, 3-D, and dynamic world that Mars is becoming has its roots in the conversations and norms of the Mapmakers' Silicon Valley surroundings.

Mars@Ames.Silicon_Valley

The Mapmakers work at NASA Ames, which is located just eight miles southeast of Stanford University. Silicon Valley is a flat coastal expanse bordered by the San Francisco Bay to the east and the Santa Clara Mountains, which are green in the summer and sandy brown in the winter, to the west. One- and three-story buildings quietly announce themselves along suburban roads with simple signs that read "Hewlett Packard," "Xerox PARC," "Apple." More recently, Yahoo, Facebook, and Google have established themselves between Highway 101 and Route 280. Between Motorola and the Computer History Museum stands NASA Ames.

The National Advisory Committee for Aeronautics established Ames as a research laboratory in 1939. Facilities were constructed on the land of Moffett Field, a military base in the town of Mountain View that was used first by the navy and later the army. In 1958 the Committee was incorporated into NASA, and the laboratory was renamed the NASA Ames Research Center (or Ames, as I will refer to it hereafter). At Ames, research on aerodynamics flourished (Hartman 1970). Under NASA's guidance, Ames branched out from aeronautics and into the life and biological sciences.[4] In the early 1960s, Ames began building ties with Bay Area scientists. By the 1970s, as Silicon Valley was making a name for itself thanks to the arrival of Steve Wozniak, Steve Jobs, and others, Ames's director Hans Mark encouraged the center's researchers to pursue collaborations and partnerships beyond NASA; Ames began conducting research projects funded by both NASA and associated universities. The center sponsored research that held benefits beyond NASA's space-oriented goals. In the 1990s, however, Ames suffered a number of criticisms brought about by a disagreement about management practices between an increasingly bureaucratic NASA headquarters and a more casually run center. Personnel and budget cuts accompanied several negative reviews of Ames. No longer able to offer competitive salaries in the midst of Silicon Valley's dotcom boom, Ames set up a career transition office to help employees find jobs at neighboring organizations.

Ames managed to thwart NASA Headquarters' plans to close it. As NASA worked to minimize redundancy, specifying the strengths of each center, Ames repositioned itself as the organization with expertise in astrobiology, aviation system safety and capacity, and, thanks to Silicon Valley's influence, information technology (IT). The late 1990s was a revival period for the center. With a campus that, like all NASA facilities, one needed a badge to enter, Ames began holding events to bring the community into its space. It held open houses to show off its robots and wind tunnels and established the NASA Research Park, in the former military buildings, where no entry badge was needed.

In 2004 NASA recommitted to human space flight with the Constellation program designed to return humans to the Moon and later Mars. With this focus on human exploration, Ames's relevance was again threatened, so its director, Scott Hubbard, created the Exploration Technology Directorate, also known as Code T (a Code is an organizational level in NASA's hierarchy nomenclature), to emphasize the role that IT would have to play in the future of human (and robotic) exploration. But Ames's budget and workforce continued to shrink. Simon "Pete" Worden, who became director in April 2006, is credited with breathing new, young life into Ames. He began recruiting young engineers who were enthusiastic about space exploration. He brought in Chris Kemp, who had a record for launching successful Internet startups, as director of strategic business development. Kemp was later promoted to chief information officer at Ames in 2008 and chief technology officer of IT for all of NASA in 2010. On Worden's wishes, Kemp created several Space Act agreements with nearby companies, including Google and Microsoft.[5] A New York Times journalist observed in 2010, shortly after my own arrival in the Bay Area, that the partnership between Google and NASA was "making Mountain View a stop along the virtual route to Mars" (Vance 2010).

During the 2004 creation of Code T and the focus on IT research, one small group within Code T, the Intelligent Mechanisms Group, was renamed the Intelligent Robotics Group (IRG; see figs. 2.1a and b). The IRG logo depicts a robotic arm reaching out to touch the extended finger of a human arm from above (fig. 2.1a). The allusion to Michelangelo's Sistine Chapel reconfigures the human as the divine being responsible for the creation of intelligent, if inferior, robots. The logo captures the primary goal of this group, which is to develop tools to assist the collaboration of humans and robots in planetary exploration. Terry Fong, who had received his PhD

2.1 The Intelligent Robotics Group: (a, left) logo; (b, below) staff and summer interns (author included) in front of Building 269, summer of 2010. Image credits: NASA/T. Fong.

in robotics, was the group's director when I arrived in February 2010 to work with the Mapmakers, a group within IRG.

On my first day, in order to introduce me to the situation of IRG within Ames, Fong told me that there are 2,500 employees there, either civil servants or contractors. Code TI, the Intelligent Systems Division, which is a branch of Code T, has 250 employees; IRG, nested within Code TI, has about 25. As one member casually described it, IRG is a "weird and diverse group of engineers, scientists, and planetary scientists who come from all walks of life." Despite its relatively small size, IRG tackles many different projects and collaborates with other groups at Ames, at other NASA centers, and in Silicon Valley. They engineer planetary rovers, work on remote sensing software, and find other ways to leverage knowledge in computer vision and robotics for planetary science. Fong explained that the common thread that unifies IRG's diverse array of projects is that they are all about bettering and facilitating exploration.

The Mapmakers work on some of IRG's other projects but spend most of their time on mapping and photogrammetry work. The Mapmakers' goal is to create more intuitive interfaces with which scientists and members of the public can explore NASA's planetary data sets. They are a young group; some are fresh out of college, some have master's degrees, and a couple of them are seasoned PhDs (though still young by NASA standards). This reflects IRG's overall demographic. Most of the Mapmakers were recruited during the Worden era when Worden was targeting Silicon Valley engineers. The Mapmakers are primarily trained in computer science and came to NASA because of an amateur enthusiasm for outer space. One Mapmaker, who was trained in cognitive psychology and computer science, recalled the denim jacket he had when he was ten that featured a big NASA logo on the sleeve. He had always wanted to work for NASA and was delighted when he saw an IRG job posting that fit his skill set. His attraction to NASA was more because of space flight than astronomy, but due to his work with planetary data he is now becoming interested in space science.

In this sense, the Mapmakers are quite different from the planetary scientists discussed in other chapters. Though all the communities I discuss are processing and interpreting data from planets, the Mapmakers are less focused on knowledge creation and more on knowledge dissemination. I am writing the Mapmakers into the community of planetary science, but they have a somewhat outsider status. They attend several of the large planetary science conferences in order to provide other planetary scientists with information about their mapping tools. For example, at the Forty-First Lunar and Planetary Science Conference, held in 2010, the Mapmakers had several posters displaying their latest lunar maps. They also interacted with the planetary science community through the exhibitor showcase. Alongside booths for scientific presses and aerospace firms, the Mapmakers set up a screen to show off Mars in Google Earth and attracted conference attendees who were interested in flying around the planet. While at the booth, scientists could pick up a flyer advertising the tools created by the Mapmakers that were free to download and could be used by the scientist to produce his or her own map of Mars. This booth advertised the Mapmakers as a team that "develops software that makes it easier for scientists and engineers to publish and access Earth and planetary imagery and data via the Internet." As toolmakers, they are supporting planetary science instead of contributing directly to the research. Yet the ability to render a planet as a familiar place is one metric of success for both the Mapmakers and other planetary scientists I discuss. At MDRS planetary geologists do so through narratives. The Mapmakers do so through maps.

Working at IRG felt a bit like working at a Silicon Valley startup. Wednesday was work at home day, and on other days people were allowed to keep their work hours flexible as long as project deadlines were met. My co-workers would drift in between 10 A.M. and noon and stay until they made sufficient progress on the bug they were fixing or the code they were writing. We worked together in a shared, open office nicknamed the Pirate Lab. I worked at the only PC in the Pirate Lab, necessary for my public outreach work with Microsoft's WWT, in a sea of Macs and UNIX boxes. I sat near the other Mapmakers who worked in the lab: Jesse, Seth, and Max. Aaron often worked at home, preferring to code at night and sleep during the day. Jesse, Seth, Max, and Aaron ranged in age from early twenties to early thirties but shared a similar aesthetic. Their clothes were utilitarian: cargo pants or jeans accompanied by casual shirts (often a free T-shirt advertising a tech product or one with a joke about programming). There were often side conversations in the lab, unrelated to the current projects. Max, the youngest Mapmaker, discussed which digital camera to buy with Seth. In turn Max gave Jesse advice on high-quality yet affordable headphones. After any of us took a trip—Seth to China, Max to the Biosphere in Arizona, or I to an observatory in Chile (see chapter 4)—we gathered around a computer and shared photos. Occasionally, Max would project a funny YouTube video onto the large, imposing monitor he said he "stole" from a different IRG lab. The daily patter in the office, filled with conversations about electronics, travel, pop culture, and other minutiae of life created a collegial and casual environment.

This relaxed environment was accompanied by a passion for the projects. They all believed in their work and either saw a public benefit or were inspired by the problems they were solving such that they were willing to stay late or work at home to finish them. There was a genuine wish to share with a larger public the treasure trove of planetary images that NASA has amassed over the years. Mars in particular has been so thoroughly documented that to capture the plentitude of image data is a programming challenge. In leveraging Silicon Valley tech know-how and creating a seamless and rich map of Mars, the Mapmakers are producing what they hope to be a democratized, 3-D and thus immersive, and dynamic experience of Mars that will allow anyone to explore the Red Planet without leaving home.

Democratized Mars

IRG projects strive to facilitate exploration not only for NASA scientists, but for the taxpayers on whom NASA depends. The millions of images NASA has taken of Mars, from the images returned by the Mariner and Viking missions in the 1970s to the higher resolution photographs that today's Mars scientists scrutinize, are all publicly available and archived online in the Planetary Data System.[6] However, the website is difficult to navigate, making it a challenge to find the spectacular images of Mars and other planets that often accompany press releases. The search tool requires the input of filtering criteria like latitudes and longitudes and orbit numbers. The website is not visually browsable but only searchable by a knowledgeable user with a specific target in mind. The Mapmakers perceive the necessity of skill to explore planetary images as a problem. They offer the map as a solution. The map is an attempt to privilege no user, to allow all equal access to NASA's wealth of planetary imagery. As the Mapmakers see it, the map democratizes Mars.

I borrow the word "democracy" from the Mapmakers. They invoked this term synonymously with the idea of "openness," a political stance they shared with fellow computer scientists involved in the free software and open source movements. In this context, "democracy" means the wide access to data and tools previously available only to the elite (i.e., NASA scientists). In a similar vein, other Web 2.0 cartographic tools have been described as democratizing (Butler 2006; Gartner 2009).[7] However, when democracy is invoked to describe mapping and exploring, it must also be considered alongside the longer cartographic history coupled tightly with western imperialism. Though the Mapmakers do not intend this association, it nonetheless draws attention to how these maps not only open up data but also enroll users in NASA's unquestioned vision of solar system exploration. As I have already implied, these maps therefore propagate a very specific planetary imagination, one that is not imperialist per se but is nonetheless in the service of a government entity's mission.

The Mapmakers are passionate and enthusiastic about making NASA products widely and easily accessible. They see their work as democratizing Mars in two ways, both by making NASA data more easily accessible and by making their cartographic tools open source. Not only should everyone be able to see the map of Mars but also everyone should be able to make his or her own map. With a map in hand, anyone can explore an unknown world.

Open Source

Aaron, who introduced me to the Mapmakers, came to IRG in August 2005 in response to a job announcement Fong had circulated on a number of computer science–related lists. He had graduated from MIT months earlier with a master's degree in computer science and had moved out to the Bay Area in search of a job. He was hired to work with a senior researcher, Kenny, on processing images from the stereoscopic data generated by the Mars Observer Camera (MOC). Kenny had already worked on the software that takes stereo pairs from ground-based rovers, like Pathfinder, to create terrain models.[8] The manufacturer of MOC, Michael Malin of Malin Space Science Systems, wanted to know if the same kind of process could be applied to images taken by his camera from orbit. Shortly after Aaron arrived at Ames, he met Steve, a planetary scientist who was a postdoc at Ames but not part of IRG. Aaron recalled that Steve attended a talk he was giving about the new project to create a "stereo pipeline," which would be a software program that takes raw data and autonomously produces a 3-D image. Steve was frustrated that scientists did not have cheap and reliable ways to make terrain models, one of the applications of 3-D images. Aaron described how he and Steve "had this notion that we wanted to build this automated tool so we could build a lot more models and democratize the whole process and get the data out there to the people so they could have access to it and hopefully, obviously, make more discoveries and whatnot." Aaron drew an "obvious" connection between democratization and discovery. Steve, who eventually became an integral part of the Mapmakers, corroborated this first encounter and the original goal of creating "something that built topography and was free for everybody to use."

This conversation about an open source stereo pipeline began in 2005, and by 2009 Aaron and Steve, aided by two other Mapmakers, Max and Seth, as well as an intern, Evan, released the first alpha version of Stereo Pipeline: "The Ames Stereo Pipeline: NASA's Open Source Automated Stereogrammetry Software." Stereo Pipeline is currently the only free system that runs on affordable, common hardware that can make terrain models from NASA's Planetary Data System.

Anthropologist Christopher Kelty writes about the culture of the free software community and its attendant geeks. Geeks, Kelty shows, care deeply for initiatives like free software and open source and the ideologies they represent.[9] They "use technology as a kind of argument, for a specific kind of order: they argue *about* technology, but they also argue *through* it" (Kelty 2008, 29).[10] The Mapmakers argue through open source technology. In striving to make their products and data open, they reinforce the logic that NASA data, produced through public funds, should be available and, further, available to manipulate, to all who are interested.

The Mapmakers released Stereo Pipeline as an open source product in 2009. It works alongside Vision Workbench, their other open source initiative. Vision Workbench contains basic computer vision tools that Stereo Pipeline utilizes but can be used for more applications than just the mapping of planetary surfaces. To make Vision Workbench truly "open," a task assigned to Seth (the "hacker" of the Mapmakers), required the navigation of many levels of bureaucracy, as he described it, before Headquarters would sign off on its release. With permission granted, Seth uploaded the code for Vision Workbench to GitHub (a website that hosts code, facilitates network sharing, and manages version control). When he announced this achievement at a Mapmakers meeting in May 2010, everyone applauded and showed genuine excitement for this accomplishment. Welcome to "this brave new world of open source," Aaron intoned.

For the Mapmakers, open source is, for the most part, an unquestioned good.[11] In daily conversations this community does not assess whether or not to pursue open source initiatives. This echoes Kelty's and Gabriella Coleman's ethnographic observations that openness is increasingly the "obvious" choice (specifically Coleman 2004, 510). The reason why open source is good is not only ideological but also practical. Open source allows for more fluid collaborations with those outside NASA, as there is no fear of sharing proprietary code. Ames in general and IRG specifically have been the primary instigators of the open source movement within NASA. In 2003 Ames engineer Patrick Moran wrote a technical report titled "Developing an Open Source Option for NASA Software." He outlined three benefits NASA would enjoy if it were to embrace open source: "(1) improved software development; (2) enhanced collaboration, in particular across organizational boundaries; and (3) more efficient and effective dissemination" (Moran 2003, 3). IRG tries to make most of its projects open source, but this necessitates the time-consuming process of obtaining permissions from NASA Headquarters.

As Kelty observes, since UNIX established itself as the model for developing code, "the norms of sharing have come to seem so natural to geeks" (2008, 119). The Mapmakers, channeling these norms, saw fit to make their code and data widely available so as to democratize NASA products. When I asked Jesse why it was important to make a map of Mars for both Google and Microsoft, he responded that it was all about providing the broadest possible access to the data. "My goal with these types of projects is to make it more useful to people. I want anyone who wants to, to be able to explore all the awesome data that NASA has. . . . Cool stuff should be discoverable. Folks in the general public should be able to just jump on the Internet and explore this stuff." Here, Jesse explains how democratizing data directly facilitates exploration. And just as open source is the obvious choice for software development, generating enthusiasm for exploration is the obvious purpose of these maps. Yet this argument needs to be made constantly because Mars is not obviously a territory ready for (human) exploration. Describing Mars as a site of collective exploration calls forth a broader engaged public that, ultimately, ensures funding for NASA and jobs for those who work with Mars data.

Opening up the Map

In developing Stereo Pipeline, Aaron, who had not previously known much about planetary science, found himself building "amazing 3-D models of Mars," and only his close collaborators were seeing them. To him, the next logical step was figuring out a way to share these views with as many people as possible and in general to make NASA's data more easily accessible. By this time, NASA's then CIO, Chris Kemp, had been in discussions with Google about joint projects. Google, whose headquarters are a short bike ride from Ames, had recently hired two engineers who in their previous jobs at a university had used Google software to do Mars mission planning for NASA. They were already thinking about how to expand Google maps beyond Earth, and Aaron and his colleague at Ames, Adam (who eventually went to work for Google), were the perfect partners to achieve this vision. The idea was to expand beyond Google Earth's virtual globe of Earth, to Mars. Mars in Google Earth was released on February 2, 2009, a year before my own arrival at Ames.[12] The NASA press release highlighted the democratizing attribute of this application: it "brings to everyone's desktop a high-resolution, three-dimensional view of the Red Planet." It went on to explain the application's uses for both the public and scientists, connecting these uses to the activity of exploring. More people could experience the Red Planet.

In 2009 Aaron was not only handling the partnership with Google but also working with the USGS and under a new contract with Microsoft to make a model of Mars for WWT. WorldWide Telescope, Microsoft's response to Google Earth, distinguishes itself as a virtual telescope, providing access to images of the entire universe, not just Earth, Mars, and a few other celestial objects of interest. With this influx of work, it was necessary to expand the team, and the Mapmakers quickly grew from two or three researchers to the seven who were there when I arrived.

Though most of the Mapmakers are trained as computer scientists, they see their work as extending the lineage of traditional cartographers charged with mapping unknown territories. During my first week, Aaron gave me his well-worn copy of John Noble Wilford's *The Mapmakers* as mandatory reading. Wilford describes maps "as interpretations of place" (2000, 16), emphasizing that the place comes first, the map second. He speaks even more grandiloquently about the relationship between maps and places where, in the last section of his cartographic history, he describes the radar mapping of the planet Venus. After seeing the map of Venus for the first time, one Venusian scientist recounted to Wilford, the planet took on a new character. As Wilford interprets that account, "mapping a place made it seem real for the first time" (295). Wilford realizes that there is something unique about planetary mapping; I would suggest that mapping is one way to give another planet the material reality necessary for being a place. It affords a tangible instantiation of the otherwise ephemeral planetary imagination.

Perhaps informed by Wilford's notion of mapping, the Mapmakers similarly see their practice as making Mars more concretely a place for scientists and the interested public. Aaron described to me the role of mapping in place-making: "before places, like California for example, are settled and before we really are living and being at a place … there are people who start the process of exploring those places and mapping them out." Aaron knit together the process of mapping and exploring, drawing however unconsciously on the frontier narrative so crucial to the imaginary of Mars described in the previous chapter.

Not only does this understanding position maps as early and crucial ways of knowing somewhere new; the Mapmakers also see this mode of representation as both a natural output of planetary data and an efficient and experiential way to provide access to as much information as possible. Jesse, who joined NASA and the Mapmakers in May 2009 and was coming from a job where he had used geospatial data to make maps (of Earth) for web applications, explained the task set before the Mapmakers as transforming all of the planetary data NASA by mandate makes public into a format lay users can intuitively navigate. He summarizes his motivation for such work in a suite of questions: "How can we make [planetary data] easy to find and easy to explore in an engaging way? How can we bring data from different modalities to the same user interface? How can we get all of this into a common space so people can really explore it and find—actually discover—new things through the synthesis of this data?" Importantly, working with planetary data that lends itself to being mapped makes achieving these goals within reach. "Making a map, which is what we're doing, is an easy way to take a lot of diverse data and visualize it in one place. It's easy because there's a geographic component to it and because it's natural for us to map—map as a verb—from that two- or three-dimensional model space to something that we can engage with in the real world." Yet there isn't just one way to map planetary data, and the Mapmakers' different projects show that no matter how scientific a map they strive to present, there are always technical and aesthetic choices that must be made that in turn shape the user experience.[13]

WorldWide Telescope, which Jesse was in the midst of working on during this conversation, was first released in 2008 and takes a slightly different approach to extraterrestrial mappings than Google does. WorldWide Telescope not only maps planets and satellites, but also uses other telescopic data sets to present a map of the whole universe. It is designed such that you can "fly" through the universe using either a mouse and keyboard or an Xbox (Microsoft's gaming system) controller. By stitching data sets together, there is continuity between flying around the Sloan Digital Sky

Survey, a photometric map of 35 percent of the night sky, and then zooming in on a beautiful, color-enhanced Hubble image of a single celestial object. The Mapmakers were brought into this project to create a "Mars mode" for WWT. Microsoft wanted Mars to be featured in a new release and to focus on the most recent data sets from NASA. In addition to the color base map of Mars from the Viking mission and the mosaic of images from the MOC camera, the Mapmakers and Microsoft decided to take the largest data set of Mars available and thus create the highest resolution map of Mars. This data set comes from the High Resolution Imaging Science Experiment (HiRISE).

HiRISE is mounted on the Mars Reconnaissance Orbiter and has been imaging Mars since 2006.[14] The pictures it takes can resolve objects less than a meter across. From orbit, it has taken pictures of the Phoenix lander and resolved the Mars rover tracks. It has taken pictures of individual boulders rolling down a hill. Because the resolution is so high, after four years HiRISE had only imaged 1 percent of the planet. The challenge for the Mapmakers was to take this 1 percent that is spread all over the surface, figure out a way to process the terabytes of information contained in this set, and intuitively embed these images in the global map. If accomplished, the user would not be relegated to hovering above the surface but could zoom down to a resolution such that one could see the ripples on sand dunes, tilt the perspective so as to view a landscape of the region, and feel as if he or she was standing on the ground. Though HiRISE images are open to the public, this would be the first time they were embedded in a map and thus the first time the public could intuitively explore them.[15]

One significant difference between Mars in Google Earth and Mars in WWT, and a difference that speaks to a democratic ethos, was the choice of cartographic projection. When making Mars for Google Earth, the Google engineers used the same map projection for Mars that they used for Earth: the "simple cylindrical projection." It is similar to the classic Mercator projection, in that latitudes and longitudes are parallel, but different in that they are equally spaced (in Mercator projections, parallels near the equator are closer together than those at the poles.)

All projections create distortions. As landscape architect and theorist James Corner suggests, maps do not mirror; they reshape. Different projections imply different sociopolitical structures. The Mercator projection orients toward the north and distorts near the poles, and countries in the

Northern Hemisphere are spatially favored. In contrast, Corner describes Buckminster Fuller's Dymaxion projection, which does not distort the poles and can be unfolded and reoriented in different ways so as not to statically favor one land mass over another (Corner 1999). The simple cylindrical projection used by Google Earth, based on Mercator, distorts the terrestrial—and consequently Martian—poles because, as Mapmaker Steve exaggeratedly put it in a public talk about Mars in Google Earth, on Earth "no one cares about the poles."

On Mars, however, the poles are extremely important for science, as they contain the only known deposits of water ice on the planet and experience seasonal melting. Consequently, they have been extensively imaged. WorldWide Telescope decided to implement a different projection for planetary mapping, the same projection they had used for the whole night sky. Jonathan Fay of Microsoft developed the Tessellated Octahedral Adaptive Subdivision Transform (TOAST) projection, which distorts every part of the surface a little bit (as opposed to a few parts of the surface a lot). Like the Dymaxion projection, it is based on a triangular geometry;[16] also as in the Dymaxion, the poles have the same amount of slight distortion as the rest of the surface. Mars in WWT depicts the poles with greater refinement, highlighting the importance of these features.

Democracy and the Spreading of the Planetary Imagination

When the Mapmakers use the word "democracy" to describe their work, they are referencing how it opens up new experiences for both scientists (in the case of cartographic tools) and nonscientists (in the case of the maps themselves). This claim of democratization, which other Web 2.0 cartographies also make, comes with several caveats (see Haklay 2013) that, during my time with the Mapmakers, were only occasionally discussed. Not only are there barriers to Internet access but also there are knowledge barriers that limit participation to the few who have adequate technical skills to truly interact with novel mapping software. Aaron admitted that Stereo Pipeline, though well-documented, is not particularly user-friendly and remains a niche and largely unpublicized application for planetary scientists. Even at Ames, when I asked planetary scientists in other research groups how they made terrain models, only those with personal connections to the Mapmakers knew of Stereo Pipeline. Though these tools promise democratization, without users they do not yet fulfill Aaron and Steve's original vision of making free and easy Mars models accessible to everyone.

Although the open source initiatives are incredibly important for the Mapmakers' professional identities, their work has greater reach through the maps in Google Earth and WWT. But it is in these maps where this desire for democracy finds an interesting coupling with NASA's imperial vision. When Microsoft and the Mapmakers released their map in July 2010, NASA issued a press release in which Ames director Pete Worden asserted, "Our hope is that this inspires the next generation of explorers to continue the scientific discovery process."[17] Worden alludes to a participatory framework, in which allowing access to science data at a young age will perhaps encourage more people to pursue science careers. Making the map of Mars widely accessible serves not only to educate and inspire but also to support NASA's mission. Again, there is nothing necessarily problematic with these associations, but democratization is, by definition, political, and in this case NASA has found a platform for promoting the excitement implicit in scientists' planetary imagination of Mars. These maps assume and thus disseminate an inherent worthwhileness in studying other planets.

The metaphor of exploration is given further credence in the way Mars is presented in these maps. Rendering the surface in 3-D, the second of the three aspects that I argue exemplify today's Mars, gives the user a sense of immersion and a feeling that one can stand on the surface and get a sense of the place. The legacy of offering the public a 3-D glimpse of Mars began in NASA Ames's backyard.

Three-Dimensional Mars

Robots have been NASA's exploratory emissaries to Mars since the 1970s. Whereas *Mariner* 9 sent back orbital images of the planet in 1971, the Viking landers descended to the surface in 1976 and returned the first landscapes of the Red Planet. Landscapes are emplaced vistas, viewable only when standing on, not orbiting above, a planet. As I have shown, landscapes spur an imagination of being on a surface elsewhere. In an episode of the original 1980 television series *Cosmos*, Carl Sagan recalled his reaction to the first photographs from the Viking lander: "This was not an alien world, I thought. I knew places like this in Colorado and Arizona and Nevada. ... [This was] as natural and un-self-conscious as any landscape on Earth.

Mars was a place." Sagan not only coupled the idea of landscape to place; he also related it to his experience on Earth. Mars inched away from the alien toward the familiar.

Images from the Viking mission were also the first to portray Mars in three dimensions, producing both 3-D landscapes and orbital views. Since then, depicting Mars in 3-D has been considered both an exciting tool for public outreach and a necessary representation for scientists studying Mars. From the early work by Stanford professor Elliott Levinthal, who popularized the 3-D view of Mars taken during Viking, to the present work the Mapmakers are doing in automating and expediting 3-D models of Mars, NASA Ames has always been a part of the 3-D initiative. Today's textural models are primary deliverables for Microsoft and Google. They are implicitly prioritized because rendering Mars as a 3-D object affords a certain intimacy with the planet. 3-D technologies have been described as immersive (Griffiths 2008), claiming to offer viewers a sense of "being there."

Despite how "natural" 3-D makes the surface of Mars look, 3-D is not easily come by. Just as a contradiction persists between democracy and an imperial sensibility toward exploration, another contradiction exists between the realism viewers claim to gain by looking at 3-D and the fact that 3-D is a product of meticulous social and technical engineering. Appreciating the extent to which the Mapmakers have concealed the challenges of producing 3-D models requires looking back at the circulation of the first 3-D images returned from the Viking landers. Whereas today's maps come close to collapsing the distinction between the image and reality, thus making Mars a place one can feel that one can experience and explore, earlier attempts were not entirely able to overcome the illusory quality of the third dimension.

Mars in 3-D

On January 15, 1979, five hundred scientists attending the Second International Colloquium on Mars gathered in an auditorium at the California Institute of Technology to watch the premier of a film, *Mars in 3-D: Images from the Viking Mission*.[18] Audience members received stereo glasses (and were assured that they would fit comfortably over their prescription spectacles) for viewing. The lights dimmed, and the opening chord of a synthesized soundtrack reverberated throughout the room as the first movie frame instructed the scientists on the best way to experience the illusion of the 3-D images. After the NASA logo was displayed against a startlingly red background, the title and credits of the film were superimposed over a series of stills, which transported the viewer ever closer to Mars. First displayed was Mars as seen from Earth, then Mars as a crescent during the Viking orbiter's approach, and finally a 3-D image of a canyon on the surface of Mars.

The narration of the film began with a shot of writer-director-scientist Elliott Levinthal, dressed in a white blazer and black tie, standing in front of a picture of the Martian surface. The scene was monoscopic, he told the audience, as "there is nothing exciting or flattering about pictures of your narrator in stereo." What he is going to present in this film is "the three-dimensional character of the surface of Mars as revealed by the Viking cameras." As Levinthal described it, 3-D was a technology used not to enhance but to accurately represent the real. The film proceeded in three parts. In the first part Levinthal walked the audience through 3-D images taken by the Viking orbiter. The camera zoomed and panned across craters, what looked like dry river beds, and the mythically named Olympus Mons to give the illusion of flying over the surface.

In the second part of the film Levinthal introduced the viewer to the Viking lander. Stereo images and video from the lander's test center at NASA's Jet Propulsion Laboratory depicted a full-scale test spacecraft perched in front of an artist's conception of the surface of Mars that was painted before Viking landed. Levinthal explained the hardware, and when he described the collecting arm, it seemed to slowly extend toward the crowd. He instructed the audience: "move your head slowly from side to side and you will observe an interesting effect." An article describing the premier of the film reported: "The long arm of the Viking Lander Surface Sampler reached out dramatically, closer and closer to each one in the audience. It seemed as though one could put out one's hand and touch it. If one moved from side to side the groping shovel followed one's every move. The more than five hundred spectators burst into applause" (Nicholson 1979).

The audience then "traveled" to the surface of Mars, and the film panned across a panorama of the landscape taken by the first Viking lander. The cameras on it were separated by nearly one meter, which made it difficult to experience the effect of 3-D. Levinthal had to coach the audience on how to achieve the proper illusion: "Concentrate first on the horizon. The prominent rock on the ridge is an ideal starting point. Relax your eyes until you begin to see the stereoscopic effect. After achieving fusion, slowly move your eyes along the horizon. Now shift your attention gradually to the foreground.... Many people have difficulty fusing the bottom half of the image. Don't feel left out if you can't see the nearest rocks. Enjoy the more distant vistas along the horizon as we take you on a three-dimensional journey over the surface of Mars."

In addition to this movie, stereo pairs from Viking were circulated in a book NASA published, *The Martian Landscape*. At the back of the book was a pocket that contained a cardboard stereoscope that a reader could place atop the pairs in order to see the illusion of 3-D. In the text preceding the pairs, the author gives guidance similar to Levinthal's. He also comments on the effect of fusing a stereo pair: "There is, however, one fairly reliable guide to the viewer's success. If, as he peers through the stereoscope, you ask him if he sees the third dimension and he responds noncommittally 'yes,' then you know he has not. Wait a few minutes and you will hear an exclamation of surprise and wonder. Then you know he has seen it. The effect is so unusual, literally drawing you into the scene, that very few people come upon it without excitement" (Mutch 1978, 145). In both accounts, the 3-D encounter with Mars takes effort on the part of the viewer. Yet still, Mars becomes so real, "you may get the impression you can actually step from your chair and onto the surface" (145). Even with these technological hurdles, Levinthal and the author of the foregoing passage cast 3-D as a technique that brought the viewer into the scene. Mars was not a flat surface but was so textured that a person wearing 3-D glasses could feel that he or she was standing on the surface. In a foreword to *The Martian Landscape*, Noel Hinners, NASA's associate administrator for space science, wrote: "Thanks to [the Viking Team], you are there." Not 'it feels like you are there,' but you *are* there. But of course you are still in the movie theater or your living room. Just as 3-D is an illusion, so too is this sense of the real.

Mars in 3-D was the pet project of Levinthal, who worked in the Genetics Department of the Stanford School of Medicine with the department's founder, Nobel laureate Joshua Lederberg. Lederberg founded his Instrument Research Laboratory in December 1959 with seed money from the Rockefeller Center. In April 1960, NASA began funding the Laboratory to encourage Lederberg and his associates to pursue the developing field of exobiology, a term coined by Lederberg.[19] After more than a decade of working in industry, Levinthal joined Lederberg at the Laboratory in 1961.

Their first NASA-commissioned project was the design and prototyping of a self-contained apparatus, the Multivator, intended as a biochemical laboratory to measure samples of atmospheric dust or soil for signs of life on Mars.

Levinthal worked on several more NASA projects, including the Mariner 9 Photo Interpretation Team, and he served as the deputy team leader on the Viking Lander Imaging Science Team. Mariner 9 was the first craft to orbit Mars and imaged approximately 85 percent of the Martian surface during its nearly twelve-month mission duration. This mission changed the geographical understanding of Mars. While earlier flyby Mariner missions had captured images of what looked like a crater-riddled, almost Moonlike surface, photographs from Mariner 9 revealed a planet with volcanoes, canyons, and many more geologically interesting features than originally thought. For the Viking missions, NASA launched two payloads in 1975. The two landers both made it safely to the Martian surface, and the orbiter imaged the surface at a higher resolution than Mariner 9. Viking was also the first mission to collect stereo pairs of images and thus the first capable of showing Mars in 3-D.

Viewing images in 3-D had been a source of entertainment for more than a century before the Viking mission. Starting in the 1820s, vision researchers devised several mechanisms for creating the illusion of three dimensions using stereo pairs. Art historian Jonathan Crary describes the stereoscope as a significant development in visual representation because it did away with a set point of view, changing the relationship between the observer and the observed. This is why, a century after the stereoscope first appeared, audiences marveled at the disconnection between themselves and what they viewed. The most enthusiastic response to Mars in 3-D came from the moment when the observer bobbed his or her head around to see the extended collecting arm from a different perspective. The object did not appear static, nor was the viewer's relationship to it fixed. In the mid-nineteenth century, the stereoscopic effect was greeted with enthusiasm because it felt more real than traditional painting or nonstereoscopic photography. As Helmholtz wrote in the 1850s, "these stereoscopic photographs are so true to nature and so lifelike in their portrayal of material things, that after viewing such a picture ... we get the impression, when we actually do see the object, that we have already seen it before and are more or less familiar with it" (quoted in Crary 1990, 124).

This correlation between "realism" and 3-D remains its primary appeal, despite the artifactory mediation needed for viewing. In a press release containing information about a screening of *Mars in 3-D* at Stanford's Dinkelspiel Auditorium on April 28, 1979, the film was described as portraying "ridges, outcrops, drifts, and craters of the Martian terrain" in "vivid, realistic detail."[20] Moreover, the film was designed to offer the public access to the way scientists were seeing Mars. The film was made on Levinthal's own accord, with support but not funding from NASA (who were producing their own public outreach film meant to encompass the whole Viking mission, not just the stereo imagery). Levinthal tried to get NASA involved numerous times, writing on March 2, 1979, to Geoffrey Briggs at NASA Headquarters to report that two thousand people had attended the screening at Stanford.[21] In late September 1980, Levinthal explicitly asked NASA to take over the production and distribution costs. Bryon Morgan, the chief of motion picture productions at NASA, responded that NASA is not in a position to cover them. "You are to be congratulated for having done a fine job. NASA is not to be congratulated for the way they have buried it."[22] Ultimately, the Planetary Society, Carl Sagan's space advocacy enterprise, took over the film's marketing and distribution starting in 1982.

One of the reasons NASA declined to take on *Mars in 3-D* was the budgetary uncertainty of the post-Apollo period. The San Francisco section of the American Astronautical Society went so far as to establish the Viking Fund, which hoped to raise $1 million by July 1980 to donate to NASA so as to continue the study of the Red Planet.[23] In an effort to reach this goal, the Viking Fund held a screening of *Mars in 3-D* at San Jose State University and requested donations of $3 ($2 for students) to attend. In a letter from the audiovisual coordinator of the San Francisco Section of the American Astronautical Society to a skeptical Levinthal, Bill Copeland explained that the purpose of the Viking Fund was not to undermine NASA but "to demonstrate support for NASA and space activities in general." He continued, "I personally have become involved with this and other space activities to open men and women to the awareness that we are not doomed to a dying planet. The intelligence and spirit of man can expand into infinite space." Copeland enclosed materials to familiarize Levinthal with the Viking Fund's mission. These stressed the role of the individual in making Mars science happen, inviting the public to "explore Mars in depth" and join in the "do-it-yourself" initiative.[24]

During the 1960s and 1970s, the images taken by the space program reshaped the planetary imagination of both other planets and Earth. The Apollo missions photographed Earth from space, and for the first time it was legible as a planet (see Poole 2010). Viking photographed Mars from the surface, and for the first time it was legible as a landscape. Both of these technological achievements recast these respective planets' placehood. As Copeland's letter expresses, he and children of the environmental movement, animated by the photograph of a fragile blue marble floating in a black void, were concerned that Earth was a dying planet. The environmental movement was one of many manifestations of a technopolitical sea change that occurred in post-Apollo America (see Moy 2004; Turner 2006). In the ensuing decades, communication technology further changed how a large number of people experience and imagine Earth, with an increasing emphasis on the global and the planetary. When the Mars program relaunched in the 1990s, the focus was, as it has continued to be, on the local.[25] NASA's rovers and landers became proxy explorers of the Red Planet's surface; the ones launching in the 2000s (*Spirit*, *Opportunity*, and *Curiosity*) even had Twitter feeds to serve as travelogues for enthusiasts back home. One technology that makes Mars local, that enlivens the spatial imagination, continues to be 3-D.

Making Mars 3-D

Since the Viking mission, 3-D has been a standard output for Mars imaging systems. Engineers improved the optics, switching from television to digital imaging and placing the lenses of stereo cameras closer together to make fusing stereo pairs easier for the viewer. Producing 3-D anaglyph images of Mars, like those presented in *Mars in 3-D*, remains a relatively simple task. Transforming these images into regional models and global 3-D maps for Microsoft and Google is the complex technical problem the Mapmakers work on.

Though they work on the problem of 3-D, the Mapmakers very rarely look at 3-D images. Instead of spending research time looking at the Mars surface, the Mapmakers look at screens that are filled with lines of code. While working on the WWT project, I was the only member of the team regularly looking at how representations of Mars appeared in 3-D. I did not realize until several months into the project that some of the Mapmakers had not seen the maps they were producing for Microsoft. They had to crowd around my Windows station to get their first glimpse of the 3-D Mars they coded for WWT.

The Mapmakers are interested in the challenge of producing high-resolution and terrain model maps for a variety of reasons. Max, who has done extensive terrain modeling of the Moon using data from both Apollo and more recent missions, had the toughest time articulating what it means for him to work on 3-D modeling. When asked about working with lunar data, he explained what that work is like by describing the kind of processing one needed to do with the images in order to clean them up. When he thinks of lunar and planetary images, he is not imagining the final product that the scientists often work with. In order to do his job—create his models—he needs to understand the process behind the final images. The mission team often does a preliminary cleaning of the data before public release. Max told me that the processing these scientists do to "dewarp" the image interferes with the geometry of the way he produces 3-D models. He has to do a lot of "book work" (calculations) to make the processed image and his code work together.

At the end of months of coding and days of processing, Max has on several occasions been the first person to see a 3-D model of certain areas of the Moon; he has been the first person to see the Moon from a new perspective. When I asked him what that experience is like, and whether it helps him understand the Moon better, geographically speaking, he responded, "No, I'm just like, it's pretty! That's what I do, I make pretty things. And I like to stitch them together and make even bigger pretty things." He got quiet before continuing:

> I don't know, I'm actually really proud of them. What I want to do is I want to get LMMP [Lunar Mapping and Modeling Project] finally finished because I want to print it out somehow and make a giant poster or do a canvas print and, like, hang it. I feel like I've done just as much work as an artist on anything else. I've spent an absurd amount of time [on it]. I've refined it. Even though it's trying to convey information, it took a lot of work to get there. . . . I don't know why I think it's cool, I just think 3-D's cool.

Here the aesthetics of scientific work are not hidden within practice, as Lynch and Edgerton (1988) diagnose in the case of astronomical image processing. Because 3-D is an illusion, a visual construct, creating a model is "cool" precisely because it is not simply "real" but a product of coding and image work. Max is proud of his work. 3-D is an aesthetic "coolness" that comes at the end of a challenging computer problem.

The 3-D maps the Mapmakers produce are not rendered as anaglyphs or stereographs, as were the early Viking 3-D products, for which special viewing glasses are needed. Rather, they are in a perspective-rendered view. Figures 2.2a and b show a 2-D image and a perspective-rendered 3-D image to illustrate the difference. Perspective-rendered 3-D makes the illusion of 3-D seem even more real and close at hand, as no additional equipment is necessary to experience 3-D. Crary speculates that one reason stereoscopes did not catch on in the nineteenth century (and 3-D film continues to feel like a fad, even in the current industry growth spurred by the 2009 movie *Avatar*) is because the viewing experience was not phantasmagoric enough.[26] The Frankfurt School argued that a spectacle is phantasmagoric only if the mechanisms of production are concealed (Crary 1990, 132). The bulky glasses needed for viewing 3-D remind the user of the illusion. With the perspective-rendered view, however, the sense of illusion disappears. This is the terrain of Mars. It can be quantified, studied, and explored. The cognitive distance between being there and virtually being there is diminished. No longer does there appear to be a need for imagination or effort on the part of the viewer. Mars seems to become a reality, but this reality is still always a technologically mediated illusion.

I asked Aaron to show me how images from the Planetary Data System get made into maps—to show me the machinery and to reveal the terrain model as an illusion. He decided to show me how to embed an image from MOC in Google Mars. I had installed Stereo Pipeline and ISIS (Integrated Software for Imagers and Spectrometers, the USGS image processing software that Stereo Pipeline interfaces with) on my computer. I opened up a command line, and Aaron walked me through the procedure. After downloading the "experimental data record" of an image from the Planetary Data System, the first step is to import it into ISIS. This created a .cub ("cube") file, readable by ISIS. The next step is to sync with the camera data. The experimental data record contains information on when the image was taken, and the camera model in ISIS will use that information to calculate the angle from which the image was taken. Next, Aaron told me to type

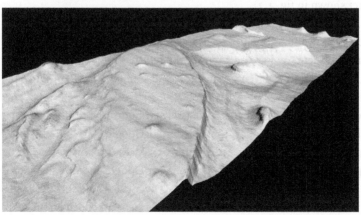

2.2 Renderings of Mars: (a, left) 2-D image; (b, below) perspective-rendered image. Image credits: NASA.

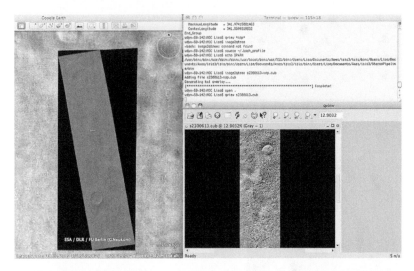

2.3 Screenshots taken during a tutorial. (upper right) Commands I typed to manipulate images; (lower right) an orthogonal image I downloaded from the Planetary Data System; (left) the same image projected and overlaid on the map of Mars in Google Earth.

the command "moccal," which calibrates the image based on information about MOC, reducing the pixel contrast to make a clearer picture. From there it is a simple command, "cam2map," that transforms the image into a map projection. In order to get this image into Google Mars, I just needed to type one more command to convert the .cub file into .kml (Google's geographic file format) and then open up the .kml file in Google Earth (final product shown in fig. 2.3).

This sequence maps a single image and mostly uses ISIS commands. Stereo Pipeline allows one to process two images taken of the same location, match them, and create a terrain model. In Stereo Pipeline, a single script can be run that, given a map-projected stereo pair, will produce the model.[27]

Though mapping one image is relatively straightforward, the daunting task that faced the Mapmakers for projects like Google Mars and WWT was to apply this sequence to the entire surface of Mars. The images had to be map projected (in two different ways, one for Google and one for Microsoft), stitched together, and then made 3-D over the whole surface. High-resolution images from MOC and HiRISE were also embedded on the map, mosaicked where there was overlap, and rendered in 3-D. To manage these tasks the Mapmakers architected a system they called the Plate File system.

Because the Mapmakers are more often than not working on lines of code than actually interacting with 3-D models of Mars, they still greet such models with excitement. During one Mapmakers meeting, Steve arrived with a box that he joked came to him "from the Internet." He explained that someone had contacted him asking if there were terrain models of the Moon from the Lunar Reconnaissance Orbiter Camera (a recent high-resolution lunar imager). Though these models have not been released, Steve had pointed the guy to the models USGS had produced using the Apollo Metric Camera. Steve opened the box to show that in thanks, the guy had sent him a model of the *Apollo 15* landing site that he had produced using a 3-D printer. Though we in the room were no strangers to 3-D, we all nonetheless fawned over the model. Being able to hold the model and turn it in our hands added yet another dimension to what it meant to interact with 3-D renderings of other places.

Doing Science in 3-D

That humans cannot physically go to Mars means that the map's user can only ever imagine being there. When scientists use 3-D, the relationship between imagining Mars and knowing Mars is interestingly blurred. The scientists with whom I spoke simultaneously praised the utility of 3-D for allowing them a "sense of place," but at the same time they relegated this spatial way of knowing to the periphery of their scientific practice. This dynamic speaks to the unique task of mapping and studying a place removed in space.

Most scientists whom I visited pointed to the 3-D glasses, either blue-red or polarized, sprawled on their desks or took them out of a drawer when I asked about how they worked with 3-D data. Others worked with 3-D in the perspective-rendered view in which Mars appears in Google Mars and WWT. Regardless of the method, 3-D renderings were ubiquitous for many Mars scientists (see also Vertesi 2015, 168–70).

How do researchers describe the scientific gain of making Mars into a three-dimensional place? Steve, the Mapmaker whose training is in planetary science and who works primarily on planetary research projects, has been working with Mars data since he was a graduate student. When he and Aaron first started working together at Ames on creating a stereo pipeline for MOC, Steve already had years of experience with the data set and was able to help Aaron navigate the data. One of the reasons he is attracted to studying Mars is because, as he said in an interview, "Mars can give us a very good sense of place. We have rovers there, we have a perspective of what it's like to be there, and we can imagine ourselves there living, working, exploring." When I asked him to describe what he meant by a sense of place, he stumbled. "Oh gosh. I can't really. I've kind of struggled with that. Like I said, I'm very visual and very three-dimensional and so part of what I enjoy doing in planetary science is not necessarily scientific but just kind of personally joyful. Specifically, what would it be like to be there or walk around and kind of be in those places and see them?" For Steve, a high-resolution image "helps, in my mind, to understand what it would be like if I stood there and I looked out. What would I see and what would that experience be like. . . . I mostly sit in this small office underground and look at images from far away across the solar system. So that kind of imagination helps get me out of the basement, so to speak. That's what I mean by [sense of place]."

Here, Steve describes his experience of Mars as a place by appealing to the role of the imagination. Mars, then, is a place both real and imagined, a place that critical geographer Edward Soja (1996) would call a Thirdspace. The difficulty Steve had in explaining what he meant by sense of place speaks to the challenge raised by Thirdspaces. In summarizing what Foucault, Lefebvre, and other spatial innovators have in common, Soja concludes: "the assertion of an alternative envisioning of spatiality . . . directly challenges (and is intended to challengingly deconstruct) all conventional modes of spatial thinking. They are not just 'other spaces' to be added on to the geographical imagination, they are also 'other than' the established ways of thinking spatially. They are meant to detonate, to deconstruct, not to be comfortably poured back into old containers" (1996, 163). Steve and the Mapmakers do not consider mapping Mars to be fundamentally different from mapping Earth and thus *are* trying to pour a new way of thinking "back into old containers." In the quote above, Steve stumbles in describing his experience with the 3-D map because, if understood as a Thirdspace, the map of Mars represents a new way of spatial thinking—one that is more difficult to comprehend than the terrestrial map.

Mars cannot be understood simply in terms of the modernist master narrative that argues for Mars as something everyone can explore and experience. Rather, experts like Steve construct what it means to experience or know Mars. Steve becomes the author of a particular way of seeing and it is incumbent on him to convince other scientists to see Mars as he does. Further, as I will later discuss, this expert way of seeing also becomes tied up with the dissemination of the planetary imagination through the Mars maps.

To help me understand his expert way of seeing Mars, Steve took me on a tour of his research and showed me how surface images and terrain models figure into his planetary research. He described various projects, showed me associated images, and explained how the numerical matrix of data that produces the images contains valuable information for his research. If most conclusions are derived through quantitative work, I asked, why do images, or at the very least 3-D visualizations, remain such a central part of research? He admitted that sometimes one does not actually need the visualized model.

> A visualization like this, kind of a 3-D visualization, happens at the beginning and at the end [of a project]. The first thing you want to do when you get a terrain model is look at it. See what it looks like. But as you note, you can't really do much with that but make screen shots to make it look pretty in a paper. You really do need to go to the numerical models and ... work with the terrain model as numbers to do quantitative science. So you do that, and then at the end you visualize it again for whatever it is you're trying to show and you put that in your paper.

Here again, the 3-D map of Mars is pushing toward a new way of knowing, which Steve resists. He insists on separating the aesthetic and qualitative visualization from the quantitative and thus scientific work. The 3-D visualizations are important precisely because they offer a sense of place, but this does not yet have a proper place in scientific work and thus is seen as outside, not a part of, the purpose of a scientific article.

Similarly, one woman I spoke with at Ames, Gloria, works with HiRISE imagery to study erosional processes on Mars but does not often use quantitative terrain models (because they take so long to build, she explained). However, she regularly looks at anaglyphs. She used to make them in Photoshop before the HiRISE team began producing them automatically. A pair of red-blue glasses sat on her desk, ready for use. When I expressed surprise that she looked at anaglyphs so regularly, she explained. "You know why? Because it gives you a sense" — she paused, searching for an appropriate description. "The lay of the land, so to speak. You get to see what's up and what's down . . . If you look at it in the anaglyph you really get a sense for how they relate to each other, these features, which you may not if you are just looking at it in 2-D. For me, I find it real helpful, just kind of a quick look of what it looks like." Despite the frequency with which Gloria used 3-D, it still remained outside her scientific practice: it was a "quick look" taken at the beginning of a project. "The lay of the land" and Steve's "sense of place" remain quasi-mystical parts of their Mars work. These steps are what help them imagine Mars as a place but do not sit comfortably next to the "scientific" Mars that their research strives to uncover.

Mars scientists have a long tradition of portraying Mars in 3-D to offer a sense of what standing on the surface would be like. This appeal to Mars as it "really" is contrasts with both the technical artifice required to produce such vistas and scientists' discomfort with the role this plays in their scientific findings. Despite a hesitancy to make central the sense of place that 3-D conjures within scientific work, for public consumption 3-D is a compelling way to transport map users to another world. For a user to appreciate such an immersive view of the planet, the Mapmakers offer guided tours within the map. These tours propagate a specific vision of Mars, one in which the Red Planet is depicted as dynamic and inviting of further study and exploration.

Dynamic Mars

Democratized, three-dimensional, and — the topic I turn to now — dynamic Mars come together in web and desktop applications like Mars in Google Earth (referred to also as Google Mars) and Microsoft's WWT. Mars is democratized in the sense that these programs "bring Mars to the people," as NASA scientist Jim Garvin once enthused to me. These applications make 3-D maps of Mars using perspective-rendered digital elevation models provided by the Mapmakers. Finally, the ways the Mapmakers and their partners choose to represent Mars in Google Mars and WWT argue that Mars is not a static object but a dynamic place. This occurs in two ways. First, tours and annotations highlight the interesting geology and processes of Mars, portraying it as an active planet. Emphasizing a past and present filled with changing land masses, flowing water, and chaotic winds implies that there is an equally energetic future, a future humans should be involved in. Second, as they cannot do with traditional paper maps, users can navigate between points in a manner that feels like flying over Mars. While working with the Mapmakers, I learned the way they wanted Mars to be dynamically experienced. Both of these dynamisms enhance the placehood of Mars by establishing a subjectivity that differs from a static map. In learning to move around Mars in a specific way, the subject's perspective shifts from that of a bird's-eye omnipotence to one of immersion. This immersed perspective mimics how scientists have come to understand Mars. Though the experience of exploring Mars feels personal, it is structured by this particular, expert way of seeing.

Google Mars and WWT do not simply serve up a blank map but heavily annotate Mars to highlight certain ways of interacting with it. For example, the Mapmakers and Google decided to include certain geographic "layers" called the Mars Gallery. Most of these layers bring Mars into a more human or terrestrial context. For example, a feature called Historic Maps layers Giovanni Schiaparelli's or Percival Lowell's hand-drawn maps from the nineteenth century over the globe of Mars. The USAF map from 1962 looks about as precise as Lowell's, suggesting to the user that visually knowing Mars is a recent phenomenon. Also in the Mars Gallery is a feature that allows the user to locate the various rovers and landers on the surface. These robots become landmarks, providing both a sense of scale and a way to navigate the planet that is linked to human history. In this sense, these maps help to "bring Mars closer," as Aaron once described it. Because the maps are infused with information about current missions and findings, they present Mars as a "place we are actively exploring." Similar to Basso's place-worlds, where elements of the past are evoked in the present landscape, maps "discover new worlds within past and present ones" (Corner 1999, 214). Maps of Mars establish the planet as a newly interesting destination by drawing on images and stories of the past and present.

These annotations are inspired by the trend in Google's other mapping products to add layers over the traditional map face. While this activity seems to make the map endlessly customizable, shedding light on the subjectivity of maps, it in fact reifies the base layer as the steady, objective surface on which new socialities can be drawn (Farman 2010). Yet even this base layer, a careful stitching together of satellite imagery, is the product of the intense technical and social work I have described (see also Wood and Fels 1992, 51).

Today's interactive maps invite the user to enter and manipulate the map. Writing before the rise of digital cartography, Michel de Certeau (1988) wrote about the changing status of the subject within maps. In *The Practice of Everyday Life* he explained the relationship between the map and the tour, describing how premodern maps used to contain itineraries (e.g., pilgrimages, stops, travel times). "The tour to be made is predominant in them." However, with respect to the modern navigational map, "in the course of the period marked by the birth of modern scientific discourse . . . the map has slowly disengaged itself from the itineraries that were the condition of its possibility" (120). Scientific mapping excluded the elements of the tour in a quest for objectivity. In Google Maps (and Google Mars and WWT), these two elements are coming back together.

For de Certeau, tours are more intimate ways of experiencing place—they are active, while maps could only be passively viewed. Mapmakers see tours in Google Mars and WWT as a way for a user to have a more intimate experience with Mars but under the guidance of an expert. In these applications, a tour is a scripted audiovisual presentation that tells a specific story about Mars. When I joined the Mapmakers in February 2010, they were hard at work on Mars for WWT. I was assigned the task of producing two tours that would be launched with the next release. Originally, Mars mode in WWT was to go live in early March, but the complexity of this project led to delays on both Microsoft's and NASA's sides. The final product was publicly released on July 12, 2010.

Chris Kemp (the person responsible for the Microsoft partnership) had already arranged for Goddard's Jim Garvin, former chief scientist of NASA, to author one tour.[28] For the second tour, I approached Carol Stoker, who, when not at MDRS or on other fieldwork trips, worked a few buildings away from the Mapmakers at Ames. As experts, the scientists offered a specific perspective, a specific way of seeing Mars, that argued for Mars as a dynamic and scientifically interesting place. These tours, then, are not a subversive experience of Mars as de Certeau might desire but are a reassertion of the scientist's role in preserving Mars as an interesting, scientific place.

Garvin was enthusiastic about his tour from the start. He came to our first teleconference with several ideas for a tour that would get people excited about Mars. His first idea was to build a tour based on a white paper he had authored that detailed future landing sites for a human mission to Mars. Another possibility he suggested would be a tour highlighting what he called "exotic Mars." He explained that we forget how alien Mars is because we so often try to understand it through terrestrial analogies. This tour would review spots on Mars that are very "nonterrestrial." We decided to pursue the first idea. Since these tours are meant to be broadly educational, Aaron and I decided that this tour should begin with a geologic history of Mars and then conclude with the possible human landing sites.

The script I wrote for this tour alluded to a changing Mars. I discussed how each era of Mars—Noachian, Hesperian, and the present Amazonian—was characterized by different geologic processes.[29] Volcanic activity and meteor impacts shaped the surface of the Mars that today's satellites image. I ended the section of the tour that discussed the different eras of Mars by writing: "The remaining water dried up but volcanic activity continued, and the Mars we know settled into existence." Though I had written about Mars's past with dynamism, this last sentence suggested that Mars now is not as exciting as it was in the past. When Garvin edited this script, he made a few changes to this section so as to destabilize the sedate Mars I presented. He included that the water "mostly" dried up and the Mars we "think we" know settled into existence. He added an additional sentence: "However, Mars is a dynamic world, with dramatic climate upheaval and we have barely begun to 'read its textbooks' even over the past millions of years."[30] Garvin wanted to show that Mars is not a solved mystery but an evolving world in need of more study.

Stoker's tour marked Mars as dynamic in a different way. The tour we put together was based on her experience as a coinvestigator on the Phoenix lander. She had been in charge of assessing the habitability of the Phoenix landing site. Habitability on Mars is defined as an area "capable of supporting living organisms with capabilities similar to terrestrial microbes" (Stoker et al. 2010). A site is deemed habitable if there is evidence of present or past life. Habitability can be both a claim about the current state of the site or a description of how the site used to be. Results from Phoenix provided a "tantalizing glimpse" of a "potentially habitable" area, and the team recommended a more rigorous search for life near Phoenix's landing site in the north polar region of Mars (Stoker et al. 2010). Again, the mystery of whether or not Mars has ever hosted life remains open and should, on this model, be an inspiration for future missions.

Stoker and Garvin recorded the narration of the tours, and I created the "flyover" of the Martian map, with overlays of additional pictures and text as needed. The first tour I visualized was Stoker's. In WWT, I stitched together slides of rotating Mars, a zoom in to one region of Mars, a zoom out and back in to another, and various pans. This was the way I intuitively thought to interact with the map. I got a draft of the tour done in time for a meeting in the middle of May with Microsoft, at which Aaron was going to present the first signature tour. The night before the meeting, Aaron told me he had done some quick polishing. The Microsoft team loved what he showed them. We watched the changes he made together in the Pirate Lab, and I realized that the visual feel was completely different from what I had produced. Whereas I kept approaching the globe head-on, Aaron almost always presented Mars from a tilted vantage point, so the viewer felt like he or she was on a ship orbiting the planet. My cinematography made Mars feel static, whereas Aaron's made Mars feel dynamic and interactive. It had never even occurred to me to present Mars the way he did, such that the user had a more immersive experience. Effectively, he trained me to see Mars the way he did so that I could create the same feel for the second tour. As he said after we watched Stoker's tour together: "Really make it feel like you're flying around Mars."

New members of the Mapmakers quickly pick up this dynamic understanding of Mars. A summer intern, Evan, heard I was interested in understanding the kind of place Mars is and sought me out one day to explain how Mars shifted in his perception from a static object to a place that was filled with movement. He told me that when he had been with the Mapmakers during a previous summer, he was working with some Mars imagery, mosaicking images by hand to think through how an automated matching program might work. He was looking at two images of the same spot and realized that a dot that looked like a boulder in one image was in a different place in the next. He soon realized he was looking at one of the Mars rovers moving across the landscape. He recalled that the moment he realized what the movement was, the images suddenly took on a dimension of place. Dynamism on the surface became, for Evan, the way in to thinking about Mars as a place.

The excitement over this view of the surface can be contrasted with de Certeau's description of his somewhat perverse pleasure of viewing New York City from the top of the World Trade Center. Like the bird's-eye view, "looking down" was a vista afforded by technology. Though he found it awesome, it concealed the places he considered most important—the alleyways and in-between areas that people use to move through the city. With Mars we had the bird's-eye view first. Through these maps and high-resolution images, technology is only now affording one the view of Mars's "alleyways." The pleasure de Certeau describes in his lofted vantage point is the same pleasure that someone like Evan gets from peering at a close-up of Mars's surface. The rover tracks are the in-between places masked by the view from above.

Accentuating the local and inducing a sense of immersion was what Aaron was teaching me when he showed me how to view and display Mars for the tours. This view, as it is propagated through the tours, then becomes the standard way to experience Mars. This is a significant influence on the public understanding of Mars, and I was curious about how the Mapmakers processed that. When talking with Max, who was still working out the aesthetic worth of his maps, he tried to guess how artists justify their products. "They like to say they're changing the world, but I think they're full of it," he said cynically of artists.

"But you're kind of changing a different world," I couldn't help but point out.

"Yeah," Max agreed, "so like we're changing the perspective, we're making Mars, instead of just being some little pink dot in the sky it's now actually some place that you've somewhat explored yourself. That you were able to see that, 'Hey, I can associate this terrain with something I've seen back home,' or 'I can see erosion on Mars' and think that at some point in time it was alive." In making Mars dynamic and alive, one makes it an interesting place to explore.

Conclusion:
Situated Planetary Knowledges

When I introduced the three aspects of planetary place that these Mars maps deploy—democratic, three-dimensional, and dynamic—I indicated that they were rich with contradictions. These maps seem to usher in a personalized era of exploration, but insofar as Mars is made an object of exploration, the map is in service of NASA's goal of eventually sending humans further into space. To offer a taste of what it would be like to visit

Mars, NASA generates excitement by circulating images taken from the surface, as with the Viking missions and most recently *Curiosity*. With the Mapmaker's digital maps, users can re-create these emplaced perspectives. The Mars that most people encounter today is not from the perspective of the global. Rather, as I have discussed, the planetary is produced (by both NASA and map users) in reference to the local.

Offering a sense of the local, then, is one strategy by which the planetary comes to have meaning. This emplaced perspective seems to mirror what Donna Haraway (1988) has called situated knowledges, yet the colonial overtones of the mapping project are at odds with Haraway's analytic intention. Haraway, in fact, uses photographs of the outer planets to illustrate the disembodied view from nowhere. This view is a "god trick of seeing everything from nowhere" (581). The user of the Mars map is ideally not hovering above the planet but viewing Mars in 3-D, from the surface. But this situated planetary knowledge is in some ways a double god trick. The user imagines himself or herself as the embodied viewer of the Martian landscape, but this is a way of seeing, as I have shown, that is specific to the Mars scientist. The scientist is visible when giving the tours but becomes in a sense disembodied once the user is trained to see. This positioning is simultaneously situated and universalizing.

Whereas I take the Mapmakers' commitment to democratizing data and their use of exploration as a benign metaphor of knowledge acquisition at face value, it must also be acknowledged that if and when human space exploration beyond the Moon occurs, ideas, metaphors, and ways of seeing that today have low stakes will attain greater power. It is then pertinent to note the exclusions inherent even within this attempt at an open and broadly accessible technology. The tools of exploration being cultivated by the Mapmakers are tools accessible to an educated, English-speaking, digitally connected user. Will the attributes of those who can be virtual explorers unwittingly constrain the attributes of those who can be future Mars explorers? And what purposes or practices of exploration will be replicated?

The answers to these questions are not obvious. The best I can do here is to return to the map and emphasize how it is not a straightforward inscription but, because of the complex interplay between the local and the planetary, offers different senses of place and different understandings of Mars. Ronald Greeley and Raymond Batson in their textbook *Planetary Map-*

ping (1990) contrast the evolution of terrestrial and planetary mapmaking. They describe how terrestrial ground surveys charted local and regional terrain and point out that it was not until the advent of aerial photography and satellite imaging that cartographers produced accurate, global maps. Planetary cartographers, however, begin with the global perspective afforded by a telescope, achieving a sense of local cartography only when orbiters and, in the case of Mars, landers approach the planet. Being situated is a position made possible only by advanced technologies, the same constellation of technologies that were needed to render Earth global.

Though my attention in this chapter has been on describing the work of the Mapmakers, their positioning within Silicon Valley, and how they produce Mars as a specific kind of place, the question remains of what knowledges of the planetary are at stake (and how these might carry into the future). Aaron and Jesse shared with me their personal understandings of Mars, each offering a different relationship between the local and the global. As much as Aaron enjoys the beautiful vistas he and his colleagues create, he is much more intellectually stimulated by the way he has come to understand Mars as a planet:

> Mars is really more Earth-like than anything else in our solar system. . . . So I've begun to think of it much more as just another place that's as rich and interesting as Earth is. . . . I have a much more global appreciation for what is going on [on Mars]. . . . One thing I've often thought is that if we can understand better the process by which Mars went from that environment with water and probably some atmosphere to where it is today, it would help us to better appreciate the Earth's atmosphere and climate. And so studying Mars suddenly has a relevance to studying our own planet.

Aaron is able to draw this relationship between Mars and Earth because he has come to understand both as planetary entities existing on similar spatial and temporal scales. Mars has meaning for Earth because Aaron imagines both on a global scale.

In contrast, Jesse finds meaning in Mars by attending to the local. He recalled how he first came to know Mars, by way of the first mission that landed on Mars in his lifetime. As a teenager in the late 1990s, he consumed the panoramic photos NASA released of the surface taken by the Mars Pathfinder. He remembered: "There was this sense of almost being there. There are these incredibly high-resolution color photos from the ground on Mars that you can explore.... And that's when I kind of started to get more of the sense that this is not something on the front page of a C. S. Lewis book, a line drawing of an imaginary world. This is a real place we can go to." Unlike the Lewis frontispiece, this was not a god's-eye perspective but a situated one.

Jesse and Aaron are now responsible for providing both the global and local views of Mars. They draw on a long tradition of both terrestrial cartography and aerial photography to present Mars as a knowable and familiar place. Aside from Earth, Mars is the most extensively photographed and studied planet. But what of planets further away, even outside our solar system, that cannot be mapped? How do scientists come to understand planets for which the global, no less the local, is invisible to even the telescopically aided eye? Ways of seeing remain important for crafting an exoplanetary imagination, though in this newer scientific field researchers experiment with how the invisible can be "seen" and creatively find ways to envision being somewhere humans cannot survive and cannot even reach.

CHAPTER THREE

VISUALIZING

ALIEN WORLDS

MIT professor Sara Seager, a couple of years after she convened the conference described at the beginning of this book, is preparing to speak to a gathering of MIT alumni. At a podium in front of a dimmed banquet room in a Philadelphia hotel, she will try to teach her audience to see the invisible planets orbiting other stars to which she has devoted her life's work.

Her talk, "Exoplanets and the Search for Other Worlds," begins with something of an eye exam. A slide behind her portrays a cloud-covered sphere. Continents and oceans peek out from beneath the clouds. "What is this?" she asks. A few hesitant hands raise; an audience member volunteers, "Earth?" Seager responds that there is too much landmass for it to be Earth. What the slide depicts is an artist's representation of how an Earthlike exoplanet might look if we were ever to find one around another star.

Seager's talk is about learning how to see exoplanets. She describes how these planets are detected and why scientists may not be finding one like Earth for some time. To explain how difficult this task is, she asks the audience to imagine themselves looking toward a searchlight in the distance and trying to spot a firefly as it passes through that light. If that is not hard enough, she says, imagine that you are on the East Coast but the searchlight is in California. This is comparable to the challenge facing exoplanet astronomers, where the firefly is a planet and the spotlight the star around

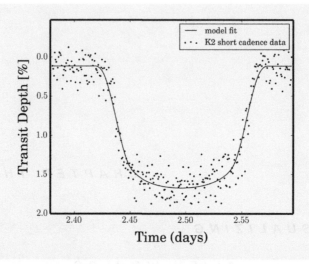

3.1 Light curve: data taken by the Kepler spacecraft in January 2014 during a performance test. The target is known exoplanet WASP-28b. The vertical axis is a measure of light emitted from this planet's star; the dip is the measured dimming of the star. This is due to the exoplanet passing in front of the star or, as Sara Seager has described it, a firefly passing in front of a searchlight. Image credit: NASA Ames Research Center/T. Barclay.

which it orbits. A slide shows a cartoon of this image. Other slides of artists' representations of exoplanets, pop-culture references, and slick NASA graphics depict these planets and their scientific and cultural import. Halfway through her talk, Seager plays an animation of a newly discovered exoplanet.[1] "The artist is imagining that we are traveling to the star," she narrates. A tiny shadow sweeps across the fiery yellow star. We zoom into that shadow and find there a treacherous, molten surface of an exoplanet. Because the planet is so close to its star, the rocky surface would continually be melting. She describes such a planet as one of many "exotic exoplanets" that bear no resemblance to planets in our own solar system.

Seager continues her talk with a special slide—just for the science-savvy MIT crowd, she explains—that depicts "real data." There is no sphere, no glowing star, just an x-axis and y-axis populated by what looks like a scatter plot (similar to the one shown in fig. 3.1). This slide is only briefly on display, but I smile with recognition, sitting in the audience, at what I know is a light curve. A few years earlier, when I was a participant observer in

Seager's research group, I learned alongside her students how to see this light curve *as* a planet, as a world.

Whereas Mars has been heavily photographed and transformed into a landscaped place, exoplanets are, to our terrestrial eyes, invisible. The light of most exoplanets are not detectable by even the most sophisticated telescopes.[2] When Seager referred to the firefly and the searchlight, she was suggesting not that we photograph the firefly but that instead we measure the momentary dimming of the more visible searchlight that is caused by the bug's shadow. Astronomers consider such slight dimming or small movement (due to the gravitational tug of a planet) of a star to be evidence of an exoplanet. Telescopes collect starlight, and astronomers translate these data into a variety of visualizations, one of which is seen in figure 3.1.[3] I will explain how trained astronomers come to see light curves and other abstract visualizations as complex planets in the same way the audience was enchanted by the artistic illustrations Seager shared in her presentation.

Scientists like Seager see these plots not only as planets but also as *worlds*. Seager ended her presentation with a slide that showed a ship Christopher Columbus could have sailed, cruising through a star-filled sky. "New Worlds," rendered in a calligraphic font, led the ship at its prow. On the slide, Seager predicted the legacy of her generation of astronomers: "[Future people] will remember us as the generation that first found the Earth-like worlds." As exoplanet astronomers await this discovery (discussed more in the next chapter) they focus on the more plentiful "exotic exoplanets." Despite bearing little resemblance to planets in our solar system, let alone Earth, scientists still imagine these planets as worlds.

Worlds, worlds, worlds. Seager, without hesitation, used this word to paint an image for her audience, but it is just as common to hear the language of worlds used during research team meetings. To describe an exoplanet as a world is to make the alien molten planet, like the one shown in Seager's talk, familiar. Martian scientists study one specific planet, but exoplanet astronomers are trying to make sense of a whole category of astronomical objects. By invoking the language of "worlds," Seager and her colleagues mark exoplanets as special, distinguishing them from stars, nebulae, and black holes. Describing exoplanets as worlds suggests that these are objects that can be not only perceived but also experienced. With enough information, one can imagine being on the surface of one of these worlds.

"World" is an imprecise word.[4] Anthropologists refer to lifeworlds and place-worlds (as I have in chapter 1) or seek to describe worldviews. These, however, are largely nonspatial uses of the term, meant to describe knowledge and cultural systems. "World," when used as a spatial descriptor of our planet, is often employed as a self-evident term and one without analytic claims. Tim Ingold (1993), in "Globes and Spheres," asks after the implications behind different imaginings of Earth. The world imagined as "globe" is for Ingold associated with a separation between human and planet, leading him to propose and favor "sphere" as an imagination that might indicate a dwelling within, not on, the planet. Ingold activates "globe" and "sphere" as analytic terms, but leaves "world," mentioned throughout the essay, as simply something that is there waiting to be described. But what if we foreground "world" and recognize its power to reach across cosmic distance and connect our world to others? Thus, to describe and maintain something as a world does (and requires) cultural work.

Exoplanet astronomers make worlds constantly, and attending to how and why they do this might open up ways "world" can be more richly used by analysts involved in conversations concerning the planetary. "World," unlike "planet," connotes an inextricable linking between Earth and humanity. It is this relationship that exoplanet astronomers leverage. "World" elevates the exoplanet above a simple scientific thing, marking it as something, someplace, that is relatable to our own experience and existence.

If the previous chapter provided an example of how a particular planetary imagination circulates beyond the scientists who create it, this chapter shows how the exoplanetary imagination, in which worlding plays a crucial role, is taught to new entrants of the field. Seager instructs her students and postdocs on both how to see data as planets and worlds and how to make others see their data as worlds. As undergraduates, students simply learn to see how data contain worlds. As training progresses, in the graduate and postgraduate stage, these worlds become more complex as astronomers try to discern more specifically the kind of place a world is. Is it a planet with an atmosphere? Does it have a liquid surface? Recounting how students acquire different ways of seeing guides two central discussions in this chapter. First, this attention to pedagogy works as a classic STS investigation of how new practices and norms are adapted and shared in an emerging scientific field. Second, I attend to how visual practice is, at the same time, a worlding practice. These two threads intertwine: one who becomes a successful exoplanet astronomer is one who becomes a successful crafter of worlds.

The Stars Align for Exoplanet Astronomy

When did astronomers begin seeing worlds beyond our solar system? Most exoplanet astronomers mark the beginning of their field as an empirical science in October 1995, when Swiss astronomers Michel Mayor and Didier Queloz announced at a conference in Florence, Italy that they had detected a planet orbiting around the star 51 Pegasi, located 50.9 light-years away in the constellation Pegasus. This planet overshadowed Alexander Wolszczan's discovery, four years earlier, of two objects orbiting a pulsar—an extremely dense radiation-emitting body formed after a star has gone supernova.

Jack Lissauer, a NASA space scientist whom I heard speak at MIT in September 2010 at a Physics Department colloquium, described why Mayor's and Queloz's detection was more significant. A planet found orbiting a pulsar is not a world. He explained, "[the pulsar discovery] didn't create much of a stir. ... The pulsar has a luminosity hundreds of times that of our Sun and most of it is in X-rays. You don't want to live around a pulsar if you are a life form anything like those in the room. ... The real excitement began when planets started being discovered around normal stars." With "normal" Lissauer is referring to our Sun's technical classification as a normal star in its stable, life-bearing middle age. This scheme presupposes a universal norm of a system capable of sustaining life; it naturalizes the pursuit of such worlds. In their discovery paper, Mayor and Queloz marked the significance of the host star this way: "51 Peg could be the first example of an extrasolar planetary system associated with a solar-type star" (Mayor and Queloz 1995, 358). Describing the star 51 Peg as "normal" and "solar-type" suggests that the worlding of an exoplanet comes not simply from its existence, but from its similarity to our system. Finding a planetary companion, as Wolszczan had done, was not enough. Finding a planet around a star like our Sun was the first step to finding a planet like Earth.

Though 51 Peg b, as the exoplanet is called, orbits a normal star, it is not a normal planet. It is not like any planet in our solar system. It is a gas giant like Jupiter, but it orbits its host star at a fraction of the distance at which Mercury orbits the Sun, coming full circle in only 4.2 days. In other words, it would be possible to argue that 51 Peg b is no more inviting to "those in the room" than planets around pulsars. The implicit excitement surrounding this discovery was that if there was one planet around a Sun-like star, there had to be many. 51 Peg b tantalizingly suggested the fulfillment of the Greek philosopher Epicurus's famous prophecy, often invoked by the exoplanet community: "there are infinite worlds both like and unlike this world of ours" (quoted in Seager and Lissauer 2010, 3).

Since the detection of 51 Peg b, more than a thousand exoplanets have been confirmed. Astronomers in the United States and Europe have started dozens of surveys, both ground- and space-based, dedicated to exoplanet detection. Astronomers primarily use two methods for detection: the radial velocity (RV) method measures a star's movement; the transit method measures its dimming. In the two decades leading up to the detection of 51 Peg b, a handful of research teams were refining the ground-based RV method to the precision necessary for exoplanetary detection (see Lemonick 1998). Radial velocity remained the dominant technique for about a decade, while a subset of astronomers perfected the now popular transit technique. Though the latter method is less precise from the ground, astronomers used transits to determine not only a planet's mass (as can be done with RV) but also its radius. From these two measurements astronomers can apply a simple mathematical calculation to derive the planet's density. On the basis of its density, astronomers can speculate on its composition. The transit method gives a better sense of what kind of place an exoplanet is.

Starting in 2006, after ground-based transit detections legitimized the technique, both the European Space Agency and later NASA launched space-based telescopes dedicated to transit searches. The most recent missions are able to detect a planet as small as Earth and, using the information gleaned from the transit, determine if it is a rocky or gaseous world. Whereas in the early days transit studies were performed as follow-ups to RV detections, today RV is often used to confirm a transiting exoplanet.

Exoplanet astronomers are appointed in physics, astronomy, astrophysics, and planetary science departments. Most senior astronomers who presently consider themselves exoplanet astronomers completed their dissertations in other astronomical areas but are now mentoring new doctoral students in the field. The number of participants at conferences dedicated to exoplanets (a few per year), averages two hundred. Of the thousands of individuals who have published a scientific article or a conference proceeding paper on exoplanets, more than 250 are authors on more than ten articles. And though this might seem a small community, published items have grown quickly since 1995. In the first years after the discovery of 51 Peg b only a handful of articles were published each year, but the number increased from a dozen or so per year around 2000 to nearly two hundred per year by 2009. Now over three hundred articles on exoplanets are commonly published every year.[5]

As this community is coalescing, so are its scientific practices. In this chapter I describe how Sara Seager trains her students and postdocs to be exoplanet astronomers. Seager was in her late thirties during my fieldwork; she is slight in her physical presence but commanding of a room when she speaks. She welcomes new visitors to her office in MIT's Green Building by ushering them to her window and showing off her sixteenth-story view of the Charles River and Boston skyline. After this, she gets to work—moving quickly from topic to topic, asking pointed questions of her visitor, and freely sharing opinions and experiences of her own life and work. In larger meetings she will interject a comment or question if she thinks the topic of conversation has wandered too far off course.

In my first meeting with Seager in the spring of 2009, she welcomed me to join her team. She put me on research projects and requested that I attend her group meetings. She also invited me to join private meetings with collaborators when she thought I might find them informative. For nine months I attended to the daily conversations and practices of Seager, her students, her MIT exoplanet colleagues, and visiting scientists. During this time I observed how learning to think about exoplanets as worlds and places, indeed how to *make* them into places, is a shared practice that reinforces the burgeoning community. Since the detection of 51 Peg b, astronomers have ceased debating whether exoplanets as a class of objects exists. However, as detected exoplanets get smaller and smaller—closer in size to Earth than Jupiter—the dimming or shifting effect they have on their stars also shrinks. Visualizations are the tools used by astronomers to convince the community of the existence of individual discoveries. Exoplanets as objects of inquiry are at times disputed. This ontological uncertainty fosters frustration during the process of professionalization. In light of this instability, anchoring exoplanets as familiar places becomes a fruitful strategy for presenting discoveries. A successful exoplanet astronomer is one who convinces the community to "see" data in the same way he or she does and to recognize that they contain a world—to persuade others to interpret the signal both as a planet and also as a viable place. In this respect, practice and identity not only shape place but also are being shaped by the pursuit of place.

Are worlds necessarily places? There is an inherent tension in these two words. Worlds are large and places more intimate. Mars was made a place in part due to images of local landscapes. Exoplanet astronomers desire the same sense of the local that is available to Mars scientists, but they are at an earlier stage in their field, and a situated view from the surface remains aspirational, just as it was to Mars scientists in the early twentieth century. At the same time, these astronomers wish for exoplanets to be as intimately known as Earth and Mars. For exoplanet astronomers, a planetary imagination helps make worlds as meaningful as an intimate, local place. This is a difficult task requiring a rich imaginary. Without high-resolution pictures of the planet, like those we have for Mars, exoplanet astronomers produce abstract representations, a sampling of which are shown in figure 3.2, that similarly strive to capture the nuances of worlds. Yet these images do not obviously represent places but are made into places through the social and technical practices around which this new scientific community has coalesced. In constructing and discussing visualizations, astronomers engage simultaneously in practices of professionalization and of place-making.

Astronomers create the different images shown in figures 3.2a through 3.2g at various stages of their research. Figure 3.2a is a light curve similar to the one Seager showed in her talk. As described above, with the light curve and a radial velocity graph (fig. 3.2c), astronomers can deduce fundamental properties of the planet: how big it is, its orbit, its density. These graphs firmly establish the existence of a planet, the first step toward making it a place. However, as astronomers begin to play with density and radius, they engage in theoretical modeling to impose limits on what the planet might be made of or what kind of atmosphere exists. Figures 3.2e through g represent different kinds of models. These images are more abstract, allow for multiple interpretations, and are perhaps harder to "see" as planets but aim at performing an important operation: making the planet into a nuanced world.

As exoplanet astronomy is a new field, I had the privilege of observing the community at a time when the techniques of seeing were still being developed. Such development is a collective task, most apparent when astronomers discuss whether or not a visualization published in a journal article represents a world. Does this way of presenting data, the astronomers would ask each other, look like a planet? More conventional representations were quickly accepted, but new forms were often questioned. World-making only happened when readers saw in the same way as the author. Perception as a collective as opposed to an individual form of cognition is echoed in the work of anthropologist Charles Goodwin in "Seeing in Depth." His account of the coordination between different kinds of scientists on an ocean vessel includes an instance when two scientists remark on an interesting feature of a visualization. To Goodwin's untrained eye, there is nothing of interest on the inscription. He notes: "Such seeing constitutes an instantiation of culture as practice" (1995, 263). Understanding how exoplanet astronomers learn to see is a window into this emerging scientific culture.

In this chapter I examine three different modes of seeing. To introduce me to the basics of exoplanet astronomy, Seager had me join two undergraduates on a project that would teach all three of us to "see with the system": to see how data contain evidence (or signals) of a star and an exoplanet, as well as the telescope and unwanted environmental interferences. The following semester I "graduated" to observing Seager train graduate students and postdocs how to "see beyond the signal" as they worked on the challenging problem of modeling atmospheres and interior compositions of exoplanets. Because so little is known about any individual exoplanet, there are few conventions for displaying such models. Seager's students were freer to experiment with visualizations. In expanding how an exoplanet could be imaged, they also expanded what aspects of the exoplanet were knowable. Other scholars have noted how new visual practices are correlated to new ways of knowing (Cambrosio, Jacobi, and Keating 1993; Holton 1998; Kemp 1997; Lynch and Woolgar 1990; Rudwick 1976). In pushing at the boundaries of what can be visually represented, exoplanet astronomers are forging a new visual culture, in Bruno Latour's sense that current practice "redefines both what it is to see, and what there is to see" (1990, 30).

Finally, I discuss how Seager and her group "see through language" in discussions accompanying the crafting or viewing of visualizations. As Jay Lemke (1998) has argued, scientific meaning is conveyed through a mixture

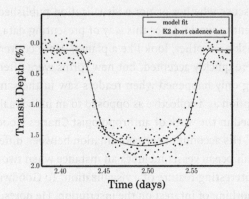

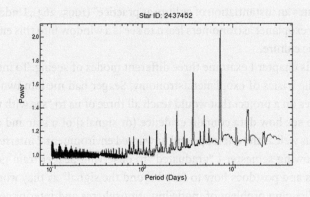

Star ID: 2437452

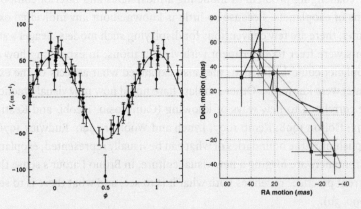

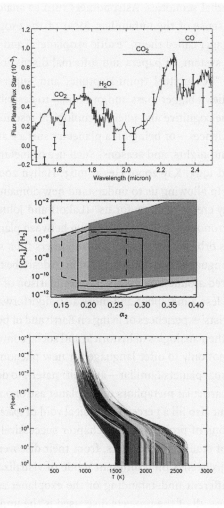

3.2 Different representations of planets. (a–d, opposite): representations of planetary detections; (e–g, above): characteristics of planetary composition. These are
not an exhaustive set of planetary representations, merely the ones I encountered
most during my research. (a) Light curve; (b) periodogram; (c) radial velocity graph;
(d) astrometry data and a predicted Keplerian orbit; (e) atmospheric spectrum;
(f) and (g) probability spaces for atmospheric composition and TP profile. Image
credits: (a) NASA Ames Research Center/T. Barclay. (b) Plavchan 2010. (c) Mayor and
Queloz 1995. Reprinted by permission of Macmillan Publishers Ltd: Nature, © 1995.
(d) Pravdo and Shaklan 2009, fig. 7. © AAS, reproduced with permission. (e) Swain
et al. 2009, fig. 3. © AAS, reproduced with permission. (f) and (g) Madhusudhan and
Seager 2009, figs. 4 and 5. © AAS, reproduced with permission.

of visual and verbal semiotics. Astronomers turn to analogies and meta-phors to make sense of the unfamiliar. Most of the exoplanets detected so far are, as Seager called them, "exotic exoplanets" and are unlike planets in our solar system. In papers and informal discussions astronomers use phrases like "hot Jupiter," "mini-Neptune," and "super-Earth" (the prefix "super" implies a bigger mass and/or radius) to describe these objects. They also use the cognitive and phenomenological resources of their own terrestrial experiences—of being on a planetary surface and experiencing weather, days and nights, and seasons. Such use of metaphor is common in science (Boyd 1993; Kay 2000; Hesse 1963; Hallyn 2000; Gentner and Jeziorski 1993). In allowing us to understand new domains of experience, "metaphors may create realities for us" (Lakoff and Johnson 1980, 156). In exoplanet astronomy, analogies are drawn between planets orbiting the Sun and planets orbiting other stars. Astronomers fill a semantic gap by appropriating language from our solar system to describe the new kinds of planets discovered around other stars. The comparison of planets to planets might seem less a metaphor and more a straightforward assumption. However, scientists' experiences of living on Earth and of being in this solar system ground their analogies in historical and social contexts. These comparisons serve not only to offer language for new phenomena but also to make new, unusual planets familiar—and, ultimately, to depict exoplanets as worlds. The particular metaphors of exoplanet astronomy not only fill a semantic void but also fill a perceived physical void of *space* with *place*.

The discussions of projects in this chapter successively work through different ways of studying exoplanets, from their discovery to their modeling. Each project relies on a different mode of visualization and a correspondingly different understanding of the exoplanet as a place. What is common across all of the projects discussed is the importance of both learning the different ways of "seeing" the data and of training readers of scientific publications to see as the author does. This indoctrinates the student into the community's shared visual practices, even as these practices are themselves taking form.[6] I will show how the cultural practices of circulating and discussing visualizations are intimately connected with the imagination of exoplanets as places. To excite the community about a particular visualization is to convince them that the image contains a world.

Seager invited me to join an undergraduate research project in the summer of 2009. I worked alongside two undergraduates, Devjit and Seth. The project concerned the Convection, Rotation, and Planetary Transits (CoRoT) space telescope, which was launched at the end of 2006 by the Centre National d'Etudes Spatiales (the French space agency). CoRoT gathers photometric data (starlight) for stellar seismology projects and for the detection of exoplanets through the transit method. When we began to work on this project, the CoRoT team had announced seven planetary detections (named CoRoT-1b to CoRoT-7b in order of discovery).[7] More significant than discovering planets, the CoRoT data set contains information on stellar variability—how the flux, or energy output of a star, changes over time. To detect a planet, a star would ideally have little variability. If a star has a lot of variability, the flux is not constant, and the star has a complex signal, making the task of isolating the signal of a planet that much more difficult. Initial reports from the CoRoT team claimed that 80 percent of stars are variable, but there was no additional information on the time scale of that variability. Seager, who was designing a space telescope array of her own to look for Earth-sized planetary transits, wanted to know how many stars might be variable on the Earth-transit time scale, as such variability would impede her search. The research project I joined was designed to answer this question in order to aid Seager in assessing the feasibility of her project, as well as teaching Devjit and Seth (and myself) what it is like to work with, as she once described it, "real data from space."[8]

Astronomers have been systematically collecting stellar data since the nineteenth century. As techniques and instruments have changed, the locus of objective measurement has transitioned from the astronomer's body (specifically his or her eye) to the increasingly mechanized telescope (see Daston and Galison 2007). This progression informs what today's astronomers mean when they talk about "seeing" stars and planets.

Beginning in the 1830s, astronomers built several apparatuses that augmented the telescope in order to give the observer's eye a frame of reference from which to make accurate observations of magnitudes. In the 1880s, astronomers developed photographic photometry as an alternative to "visual" (seeing with the eye) photometry.[9] As an editorial in *Nature* summarized it, "the eye ceases to be the actual photometer employed. For the impression on the retina we have substituted the impression recorded on the photographic film" ("Improvements in Photometry" 1895, 560). The light of a star leaves a circular impression on a photographic plate: the brighter the star, the greater the diameter. In the late 1890s, magnitude was determined based on a logarithmic relationship between the measured diameter and known magnitudes of reference stars.[10]

Throughout the first half of the twentieth century, photographic photometry was the dominant method but had many mechanical difficulties. With the invention of the photomultiplier in 1950, photographic photometry finally enjoyed expediency and success. The dominance of the photomultiplier was short-lived, as electrical engineers at Bell Labs soon invented the charge-coupled device (CCD) and astronomers began to digitize their data collection. Today CCDs have completely replaced the need for photographic film and are used as detectors for all major ground and space optical telescopes (as well as many personal cameras) (Hearnshaw 1997).

Most texts on the CCD use an analogy created by Morley Blouke and Jerome Kristian to explain its mechanical workings. As a 1992 article on the current state of CCDs in astronomy describes it: "Imagine an array of buckets covering a field. After a rainstorm, the buckets are sent by conveyor belts to a metering station where the amount of water in each bucket is measured. Then a computer would take these data and display a picture of how much rain fell on each part of the field. In a CCD system the 'raindrops' are the photons [and] the 'buckets' the pixels" (Janesick and Elliott 1992, 6). What this analogy fails to mention is that often you are interested in the raindrops coming from only one specific cloud. Yet the buckets collect rain from many clouds and perhaps even rain that dripped off an overhanging tree. Switching back to the language of astronomy (though this is still a borrowed analogy from signal processing), the raindrops from the cloud of interest are the "signal," and everything else is "noise." The skill and craft of CCD photometry is isolating the signal from the noise.

The CoRoT satellite logs stellar data on its CCD, which Centre National d'Etudes Spatiales astronomers then download. After the CoRoT team has had time to do a preliminary analysis, they release the data to the broader astronomical public. Our summer project made use of this public data set.

Devjit, Seager, and I met for the first time in early June. (Seth would join us two weeks later.) Devjit had just finished his junior year and hoped to turn this project into a publishable article to help out his astronomy graduate school prospects. Eager to impress his research advisor, he diligently took notes in a fresh lab notebook. Because Seager was trained as a theorist, she was also new to working with satellite data. At this first meeting, the three of us gathered around her computer, and Seager typed in "NStED"—for NASA Star and Exoplanet Database—to show us how to find the CoRoT data.

This database is a public repository of stellar data from various exoplanet surveys.[11] Data on this site are often already cleaned of noise from the CCD itself. Each downloadable file contains all of the observations made of a particular star. On the database, Seager pulled up one of the CoRoT stars and showed us a light curve. The task she gave Devjit was to figure out how to download the FITS file (an image file format used in astronomy) and use MATLAB (a technical computing software) to replicate this image. If he could replicate the image, then we could be confident that we were using the correct data in subsequent analyses.

The next day when Devjit and I arrived for our noon meeting with Seager, she was unexpectedly delayed on a conference call. Waiting outside her office, I noticed that a map of Earth, simply titled "The World," decorated the wall. The world or a world, I wondered.

When Seager called us in, Devjit immediately opened up his laptop, excited to show us that he had accomplished his task. On his screen were two light curves—on the left was the one we had seen the day before on the database, and on the right was an identical graph he had produced in MATLAB. Seager was excited by this proof of concept but also terribly frustrated that the data were so "noisy." Devjit had been playing with data from a star known to have a transiting planet. A light curve is rendered as flux over time—the brightness of the star over the duration of an observation. If a star's brightness does not vary, the light curve is a straight, horizontal line. Any brightening or dimming is represented by a rise or fall of this line. When a planet crosses in front of its star, it blocks out a minuscule fraction of the star's light. This shadow, the exoplanet's signature, is recognizable as a U-shaped dip in the light curve. While there were dips in the CoRoT curve we were looking at, there were also peaks and other unrecognizable shapes that made the whole graph look chaotic, not like an orderly U.

We had expected this light curve to contain only the flux from a single star. However, light curves downloaded from NStED (fig. 3.3) contain signatures of the star, the instrument, and even the Earth. Knowing how to

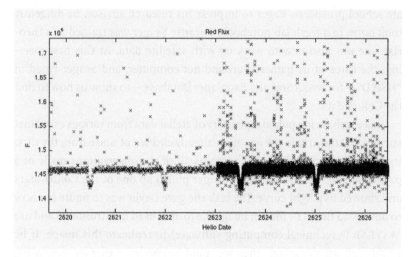

3.3 Raw CoRoT data: light curve for CoRoT-1b, the first exoplanet detected by the satellite. The periodic dips in data are transits. When it seemed likely that there was a transit, the satellite was instructed to "oversample," which is why after Helio Date (short for Heliocentric Julian Date) 2623 the curve is thicker. Image credit: author.

work with this data, how to manipulate it in such a way that it serves as a representation of the star alone, requires an intimate knowledge of how and where the CoRoT satellite operates. Team members of CoRoT try to render their apparatus transparent, providing articles and manuals detailing the effects of the satellite's orbit, Earth's interference, and attitude control fluctuations in the light curves (see for example Aigrain et al. 2009; Auvergne et al. 2009). What we learned that summer was how to see with the system in order to distinguish signal from noise.

Distinguishing "real" data from "artifacts" is a common pursuit across many sciences. Michael Lynch writes: "The possibility of artifact is an almost inevitable accompaniment of research which relies upon specialized techniques and machinery for making initially 'invisible' theoretic entities visible in documentary formats" (1985, 82). He observes that artifacts provide the analyst with an opportunity to see more clearly how images are constructed, to make visible the sometimes invisible work of scientific practice. The skill of isolating artifacts in exoplanet astronomy comes from professional experience working with astronomical photometric data and familiarity with the specific telescope. As Devjit, Seth, myself, and even

Seager were short on these attributes, we did not understand how to get from the messy downloaded data to the clean graph printed in publications. Our work progressed only because of two visitors who had prior experience working with the CoRoT data. To see exoplanets in the data, young astronomers must learn to see data as a composite of the telescope, the star, other environmental factors, and, of course, a world. Seeing the data in this way is a cultural activity that, when learned, distinguishes data that contain a place from those that do not. Seager excitedly shared the news of these visitors, both European astronomers who had published on the planet-finding capabilities of CoRoT. Richard was a new postdoc in the MIT Physics Department, and Victoria, a coinvestigator for CoRoT, was in town to work with collaborators at the Harvard-Smithsonian Center for Astrophysics. The questions Devjit and Seth asked at these meetings sought to elicit the visitors' personal experience.[12] How do they see with the system?

The explanation they gave of their experience worked across words, computer screens, and chalkboards. The heart of any astronomical system is the telescope. So Richard and Victoria both began their separate meetings with us by asking if we had seen a picture of CoRoT—did we know what CoRoT and the CCD physically looked like? Richard drew a picture of the CCD on the chalkboard, explaining how it collected starlight. Victoria, weeks later, opened a PowerPoint presentation on her MacBook and offered an image of the entire satellite and where the CCD was positioned. We learned from our visitors that to see the data properly we had to have an image of the whole system (star, satellite, Earth) as a lens for interpretation.

The data coming directly from CoRoT are noisy, Victoria confessed, and have to be properly "cleaned" or "reduced." Knowing about the instrument and its orbit allows a first-order elimination of "bad data" within a light curve. Even after these data have been removed, noise remains that impedes the detection of a planetary signal. Astronomers employ many methods to extract this signal. Continuing to draw from the language of signal processing, these methods are called filters. Richard and Victoria used different filters to work with the CoRoT data. In both meetings Seager was intensely interested in what these filters do and why these specific ones were chosen. As a theoretician, Seager rarely works with such raw data. During our meeting with Victoria, when she explained that she prefers a "five-point boxcar" filter because a former boss introduced her to it, Seager was frustrated that there was not a more systematic reason for this selection. It reaffirmed why

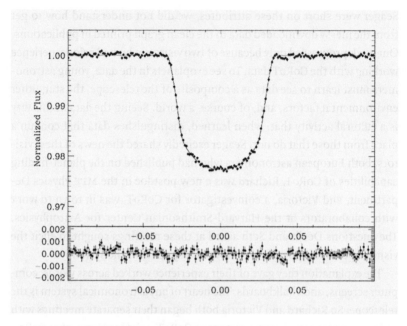

3.4 Cleaned and reduced light curve for CoRoT-1b. Image credit: P. Barge et al., (2008), fig. 1. Reproduced with permission. © ESO.

she went into theory; so much of observational astronomy is lore, Seager explained. It cannot be learned in a book.

Devjit and Seth had learned that the data had to be passed through a filter from reading papers published by the CoRoT team, but they did not know why this was a necessary step until Richard showed us before and after images on his computer. He pointed out a cosmic ray (particles captured by the Earth's atmosphere that, when they interfere with CoRoT's line of sight, cause an intensity spike unrelated to the star) in the original data. He explained that this obfuscated the transit signal and so he removed it using a moving median filter. He then went to the chalkboard and illustrated how this filter works on a simplified data set. Seager was skeptical and asked why he bothered with this step. To convince her, he showed her the afterimage of his light curve, and she exclaimed, "Oh, I can see the transit." Richard justified his method by disciplining Seager's way of seeing.

A transit of CoRoT-1b, the first planet detected by this satellite, can be "seen" in figure 3.4, which shows a published representation of the same data presented in figure 3.3. Not only has this data been cleaned of orbital and terrestrial signals and run through several filters, it has also been "phase-folded," that is, every observed transit is superimposed so as to emphasize the curve's shape and thus convince the reader that a planet passes in front of the star.

Michael Lynch and Samuel Edgerton Jr. describe the aesthetic project of producing digitally "cleaned" astronomical images. They demonstrate how astronomers, when processing their scientific images, evoke an aesthetic not of beauty but of realism. The "noisy picture" becomes "a visually coherent and naturally interpreted astronomical display" (1988, 212). Lynch and Edgerton focus on showing the role art plays in scientific practice. The description they offer of CCD processing well narrates my own encounters. However, what they leave out and what I here emphasize is how this aesthetic becomes shared—how a convention of seeing gets accepted by the broader community. The article announcing the discovery of CoRoT-1b (Barge et al. 2008) asks the reader to see the data with the system in the same way that was demonstrated in our meetings with Richard and Victoria. The article begins by describing the CoRoT instrument, its orbit, and the CCD, explaining how data is reliably ferried from space down to Earth. It goes on to enumerate the steps taken to clean and reduce the data and finally describes how the transit curve was mathematically fit to provide the best possible planet parameters. The community considers the planet CoRoT-1b a viable exoplanet because they accept the article as a successful argument that the visualized transit is constructed only from the starlight of CoRoT-1 and the effects of the surrounding system are removed.

Through the summer of research, Devjit and Seth learned how to get to know a data set. They were taught how to see with the system, which translated to knowing how to interpret light curves. To believe that there is a world in a light curve is to see and trust the methods by which they are produced. Toward the end of the project, both undergraduates were dismayed because each step of analysis took much longer to complete than anticipated and they were not able to answer the science question that motivated the research. Nor did they ever get the data to entirely look like the curves published by the CoRoT team. Seager explained several times that the process they went through was invaluable, as they were learning the general philosophy of working with telescope data. Had they just read an article describing the steps of working with the data, they would not have encountered the nuances of each step and the associated assumptions.

Devjit and Seth learned that producing a clean exoplanet signature was not a straightforward process but one that relied on detailed knowledge of the system and subjective choices, such as deciding which filter to use when. Most astronomers are aware of this complex process, so convincing fellow astronomers that a world exists often requires more than just cleaning up and phase-folding the data. About a year after my work with Devjit and Seth, the MIT exoplanet community was hotly debating whether or not a newly announced planet, Gliese 581g, really existed. The discoverer, Steven Vogt, claimed that this was the sixth planet to be discovered around its star and, significantly, was at a distance that water might exist on the surface—a habitable planet (see chapter 4). This would be the first such finding and big news for the community.

A dozen of us sat on the eighteenth floor of the Green Building. As we began discussing the Gliese 581g discovery paper, Seager suggested that we "start with whatever figure is most important and go back from there." There was no transit light curve, but a variety of other representations were meant, when viewed together, to show the five known planets and the new sixth planet. These were not conventional visualizations, and the group was not convinced—no one could "see" the planet in the data. There was confusion as to how the data were sampled and cleaned. One senior astronomer offered an alternative account of how he would have worked with the data.

Their skepticism was appropriate; a few weeks later, during an exoplanet conference in Turin, a Swiss astronomer who had access to more data than the original discoverer said that the signal was not found in the larger sample. Describing this talk to the MIT crowd, astronomer Josh Winn stated that the new data confirmed the existence of the previous planets but a "new planet did not emerge" from the data. Gliese 581g never became a world or even a planet for the exoplanet community because the visualizations and accompanying text differed from the shared cultural logics— they were not standard representations, nor were the authors of the discovery paper successful in training their readers to see with the system. Not only did a planet fail to emerge from the data, but a place the discoverers thought was covered with oceans and land was unmade.[13]

When astronomers first learn to see worlds by seeing with the system, they understand the orbit of the satellite and the journey of the starlight in order to think through "systematics," or periodic artifacts, in the data. Once an astronomer detects an exoplanet signal, the astronomer wishes to see *beyond* the signal—to consider what kind of surface, composition, or atmosphere this world contains. The two-dimensional light curves give way to planetary places as astronomers discuss, imagine, model, and speculate about being on the surface or floating in the atmosphere of these planets. The compositions of exoplanets are highly unconstrained; that is, observational data do little to limit the range of a planet's chemical and material properties. Astronomers nonetheless push these limits, as two graduate students in Seager's group demonstrated with their attempts to model the atmosphere and interior compositions of different exoplanets. The visualizations they produced are more complex and harder to "see" the planet in because in moving beyond the signal, uncertainty in the form of probabilities must be introduced into the representations. Consequently, exoplanets become less singularly understood, opening up the possibility for multiple interpretations of what kind of world a signal might contain.

With a light curve, though it takes work to prove that a dip in the data exists, once the astronomer establishes a dip, it becomes synonymous with an exoplanet. For these complex models, the exoplanet is known to exist, but what is at stake is what kind of exoplanet it is. The development of these visualizations positions the exoplanet as a place-in-the-making—a place on unsteady ground that depends on community acceptance to be maintained as such.

Atmospheres

Madhu, a graduate student of Seager who, when we met, had just finished his PhD and stayed on for a postdoc, laughed at the absurdity of his research when he explained to me that he had chosen his thesis topic, modeling exoplanet atmospheres, because he has always been drawn to unsolved problems.[14] He got to this research because he kept choosing the road not taken. He described how Robert Frost's poem informed his academic path:

At every point in life you get two choices at least, and you are to make a selection. In the initial stages of life, you start from ground zero, you are broke, you have nothing. Then, the choices you make might affect your future very drastically. . . . As you go on, the choices you make become slightly less important. At this stage, for example, if I ask "do I work on exoplanet atmospheres" or "do I work on planetary detection" . . . I'm good enough I'll be relatively successful in both. So that might have slightly less influence of where my future is. But long back, I couldn't say the same thing. . . . The summary point is how do you come up with a coherent theory, which helps you in making a decision in each of these situations, whether you have a big or small decision. The theory is that poem. . . . You look down each road as far as you can and then your gut feeling says maybe this is what I should take. It is all art at the end of the day.

While choosing his path in life is an art, his research on atmospheric composition is utterly systematic. Throughout the early 2000s, astronomers using the Spitzer and Hubble space telescopes made claims about exoplanetary atmospheres.[15] Astronomers like Madhu are interested in both the elemental composition of the atmosphere and the temperature-pressure (TP) profile. In ascending from the surface of a planet, the atmospheric pressure decreases. The TP profile is a model that gives dimension to the exoplanet.

But how, Madhu asks in his work, can you actually say with certainty that there is water or sodium or carbon dioxide on these planets? Given the scant observations, that is, the lack of a complete spectrum, how can astronomers confidently describe the atmosphere as one set of molecular combinations versus any other? How can a scientist present one TP profile or spectrum and be sure that this is the only possible configuration? Madhu's work on exoplanet atmospheres problematizes two kinds of representations. The spectrum of chemical composition and TP profile (shown in figures 3.5a and b), are related to each other through a set of equations and are solved for simultaneously. To explain the modeling problems inherent in these representations, Madhu showed me a *Nature* article (Swain, Vasisht, and Tinetti 2008) in which the authors claimed that exoplanet HD 189733b must have both water and methane because water alone does not fit the data. Madhu's point is that a wide range of water and methane concentrations and any arbitrary TP profile can fit the data. At the time of this

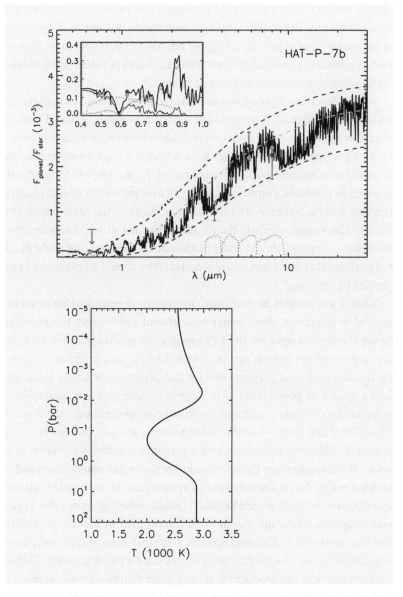

3.5 Two representations from a well-studied exoplanet: (a, top) spectrum and
(b, bottom) TP profile for HAT-P-7b. The clarity of these graphs suggests that the
spectrum and TP profile are precisely determined. Image credit: Christiansen et al.
2010, figs. 7 and 8. © AAS, reproduced with permission.

article, research teams were only running a few models and choosing a nominal best fit. The goal for Madhu's dissertation was not to suggest one or two models but to provide a range of possible models and, more important, to outline the physical constraints of the model in a statistically meaningful sense by computing millions of models.[16]

Madhu described figuring out an inroad to this problem as a depressing undertaking—after all, if he could not solve this problem he would have to start over again with a new thesis topic. Planets in our solar system have robust TP profiles because scientists have collected many measurements, so he would look to these graphs for inspiration. At one point in his graduate research he produced a figure containing all four giant solar system planets (Jupiter, Saturn, Neptune, and Uranus) and Earth in the same frame. On viewing this congruence, Madhu recalled, "an idea struck, a very creative moment. . . . Maybe I should find a mathematical function that could fit all the profiles. That itself was an absurd idea. Why should all planetary [TP] profiles be the same?"[17]

Yet this was the idea he developed, proposing an equation that could be applied to all planets, those in our solar system and beyond. Not only did he use the general equation to fit TP profiles and spectra from the data, he also presented the models not as a single line but as a probability space. He transformed what previous articles had presented as single solutions into a spread of possibilities that suggest visually that the atmospheric compositions of these planets are far from being understood. Madhu used a familiar visual form—one that astronomers can "see" planets in—and altered it slightly to accentuate how a planet is not obviously there. In a sense, this denaturalizes the relationship between the visualization and a physical reality. This is similar to the analysis offered by historian of science David Kaiser in "Stick-Figure Realism" (2000), where he writes that Feynman diagrams, a now ubiquitous visualization used by particle physicists that was developed by Richard Feynman in the mid-twentieth century, were not successful because they uniquely represented a physical reality. Rather their success was the product of, among other things, a visual resonance with more established conventions. Kaiser offers evidence of the divorce between representation and reality by showing how other theorists used Feynman diagrams to undermine the very theory Feynman used to derive these paper tools. Kaiser argues that visualizations are generative and provide "a heuristic scaffolding" for other interpretations (51). The visualizations Madhu produces with his million models method similarly build on existing visualizations to make a different claim about the physical world. In a field where small amounts of data are leveraged to make big claims, Madhu's intervention was to show that the visual clarity of spectral graphs is misleading and the known attributes of planets are much more messy and abstract (see figures 3.6a and b).

In general, this approach was well received. However, critics were quick to point out that his elegant equation for the TP profile had no physical basis. In a public talk he preempted the audience and rhetorically asked, "What is this magical pressure temperature profile?" He went on, "[it] looks like a mathematical construct, why should we trust this?" His answer was a visual argument. He displayed a slide showing how the equation fits the data we have on exoplanets and, more impressively, five solar system planets. In reference to the solar system he said, proudly, "That's real data." Representation and analogy merged as Madhu justified his way of knowing exoplanets. His equation, derived from a particular way of seeing planets in our solar system, is justified by disciplining the audience's sight to see in the same way. More persuasively, the equation illuminates faraway planets by relating them to those nearby.

Madhu is reshaping visual practice, presenting new ideas about what can (or cannot) be seen and known about exoplanets. His way of seeing beyond the signal portrays a planet not as a single entity but as a multiplicity of possible incarnations. In some sense, he is going back a step from the pervious modelers, who offered one or two interpretations of a planet. In destabilizing these singular representations, he is suggesting that these are still places-in-the-making.

Interiors

GJ 1214b made headlines in December 2009.[18] *Wired* magazine invited readers to "meet GJ 1214b, the most Earth-like planet ever found outside our Solar System" (Keim 2009). The *New York Times* warned readers: "Call it Steam World. Astronomers said Wednesday that they had discovered a planet composed mostly of water. You would not want to live there" (Overbye 2009a). As these statements make clear, the character of GJ 1214b is ambiguous; it is not singular but multiple and is interpreted by journalists as both a friendly planetary kin and a hostile world. In terms of mass and

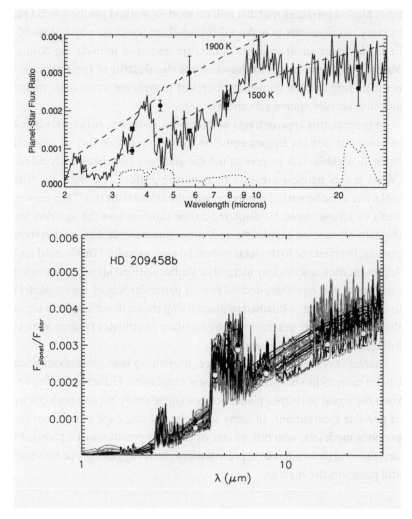

3.6 Spectra of HD 209458b. The singular top spectrum (a, top) contrasts to Madhu's multiple probability spectrum (b, bottom). Note, however, that the eclipse depths of this spectrum were significantly revised in later studies. (See Diamond-Lowe et al. 2015.) Image credits: (a) Madhusudhan and Seager 2009, fig. 9; (b) Knutson et al. 2008, fig. 3. © AAS, reproduced with permission.

radius, GJ 1214b was one of only two "super-Earths" known at the time of its detection, making it the most "Earth-like" exoplanet. The discoverer, Harvard's David Charbonneau, claimed in his discovery article that the composition was mostly water, with an atmosphere that would make the surface "inhospitable to life as we know it on Earth" (Charbonneau et al. 2009).

Is such a specific compositional claim justifiable? The planet GJ 1214b presents a puzzle in modeling planetary surfaces and interiors that is similar to the puzzle of Madhu's project.[19] How far beyond the data can astronomers go in speculating about a planet? An observer might be satisfied to offer one suggestive interpretation that fits the data, leaving it to the theorist or modeler to be more comprehensive. In September, a few months before the public discovery announcement, Charbonneau had shared his unpublished data with Seager. Seager suggested to her graduate student Jessica that this would be a good project for her to work on. As I had successfully completed my undergraduate research project participation, Seager invited me to follow the development of Jessica's research, wanting me to understand how mass and radius (quantities derived from the light curve and radial velocity measurements) are indicators of internal structure.

Jessica was in many ways the antithesis of Seager. She demurred where Seager asserted, was nervous where Seager was confident. Seager, impressed with Jessica as a physicist, was sure that she was ready for a project of this sophistication. Jessica decided to work on it while she was figuring out what her thesis research would be. This short project, hopefully only a couple of months of work, would serve to test whether there were enough data and theory at this point to work systematically through the problem of interior modeling. Consequently, Jessica's anxieties about her research trajectory and abilities as a scientist percolated during working sessions and meetings with Seager while processing the data from GJ 1214b.

"What we're doing is we're trying to interpret a new planet," Seager explained to me the first time the three of us sat down. From the transit data and the measures of the mass and radius, it seemed that the planet had to support an atmosphere, or gas layer, as Jessica referred to it. Jessica sought to explain how interior composition dictated the origin and composition of the gas. Jessica and Seager conceived of the planet as a whole, where one layer affected the others. For her project, Jessica was considering three different ways this planet could have formed based on different "primordial" material. Jessica explained these planetary possibilities to me. The first was a "mini-Neptune" with a hydrogen-helium gas layer. The second scenario was a planet that did not have a hydrogen-helium atmosphere and instead was enveloped by a water vapor layer. The last case was a super-Earth encased by a small hydrogen layer.

I was curious to know how they arrived at these three scenarios. As Madhu had shown me, there are many—seemingly infinite—possible planetary configurations. Jessica and Seager wanted to present discrete cases and chose to consider "end member" cases: planets that evolved in three different ways and therefore had three different evolutionary reasons for having a gas layer. At early meetings and in early article drafts, this clarity was not yet fully formed. Drawing again from our solar system, Jessica explained that there are rocky planets like Earth and Mars and Venus, and ice giants like Uranus and Neptune. The mass and radius of GJ 1214b fell between these two classes, suggesting other kinds of planets in this overlap region. Consequently, they chose cases based on intuition but knew that for the purposes of the article's argument, they had to present readers with a more methodical motivation behind these choices. In the final article, they justified their choice of cases with two figures illustrating the multiple possible elementary compositions for planets. They sought to convince the reader of the rationale behind their approach by using images in addition to words.

Jessica arrived frustrated to a meeting in October. After a disheartened discussion of how she was modeling the water vapor layer, she brought up a colleague's recently published article that was also on the topic of modeling interiors of super-Earths. Jessica thought she had encountered what all researchers and academics fear: that someone else has written the article one has been working on. Seager quickly reassured her, emphasizing that Jessica's article had better physics and contained a discussion about how to distinguish between different cases. Jessica was not satisfied. The other author was a postdoc and would always be a step ahead, she insisted. This should not be a concern, Seager responded, because of Jessica's proficiency at physics. But if being the best physicist was not a satisfactory solution for Jessica in terms of distinguishing herself, her other option was to embark on something completely different. Jessica had already been toying with studying planetary formation, as perhaps this would be a more novel thesis topic. The uncertainty of exoplanets as objects fueled her uncertainty about working as an exoplanet astronomer.

Frustration continued to plague Jessica's work on this project. At the start she was confident that the data would offer some hard limits, allowing her and Seager to pronounce on the amount of hydrogen in the gas layer. Two months after working with the data, after she had drafted an article, Jessica concluded that there was not enough information to constrain the problem. "Who knew it would be so degenerative?" she asked, defeated. Seager hesitated, uncharacteristically, and then bluntly observed that the project was in a precarious position; if they couldn't fully articulate the kind of place this planet was, should they scrap the idea of publishing? Jessica was stunned that her hard work might so easily be dismissed, but Seager reassured her that the null finding was not her fault. Seager and Jessica were both disappointed that there was not a bigger finding to report but decided, on advice from a colleague at MIT, to continue writing up the article and at the very least to point out that exoplanet modeling is still in its infancy and such claims of "mini-Neptunes" and "water worlds" are premature — exoplanets were still places-in-the-making. This experience made Jessica question whether studying interiors was too risky a thesis topic. As we rode down in the elevator together after one of our meetings with Seager, Jessica said that as soon as this article was submitted she was going to look for a new project, one decoupled from the worry of whether or not a specific exoplanet existed. For Jessica this would provide a sturdier platform for her professional identity.[20]

By early November, however, Jessica and Seager made the decision to keep working on this modeling project and to publish an article. Knowing that Charbonneau might announce the detection any day, Seager and Jessica worked nonstop to prepare their manuscript. Though Jessica had written a draft, Seager worked with her for several days to rewrite the article so it more clearly expressed the goals and findings of the project. Seager sees it as part of her role as an advisor to teach students how to write, a skill she says they have rarely mastered before entering graduate school. As Seager walked Jessica through the article-writing process, Seager also instructed her student in how to see planets in a certain way. They met on a Monday morning in early November for the first of several days of intense collaborative writing.[21] The first method of attacking this work was to approach it visually. Seager placed a sheet of paper on the table and suggested that she and Jessica sketch out pictures of the different end members so as to understand the physical components of each scenario. As she drew each case—narrating while she drafted an icy core and a gas envelope—she ordered them by size, illustrating a continuum of possibilities. In drawing this out, Seager created a pictorial rationalization for the selection of end member cases, which they next translated into words for the article's introduction. Seager demonstrated for Jessica how to see beyond the signal, beyond the transit light curve, toward multiple scenarios.

Ultimately, the article concluded: "we can constrain GJ 1214b's composition but we cannot infer its unique true composition." The article does suggest, however, that Charbonneau's claim that there was liquid water on the planet's surface was unlikely to be true. To model exoplanets is to see them as several. It is to resist the desire to ascribe a singular makeup, to present them as unambiguous worlds. Seager, Jessica, and Madhu played with the ontology of exoplanets. They *enacted*, to borrow from Annemarie Mol (2002), an exoplanet as multiple. Unlike in Mol's case, where atherosclerosis is made multiple through the practice of different actors, the same actors—in this case Jessica and Seager—are making exoplanets multiple. To do so, however, necessarily introduces uncertainty not only into the ontology of any particular exoplanet but also into the epistemology of exoplanet astronomy as a field. If what is compelling about an exoplanet is that it is a world, as the placehood of the exoplanet destabilizes, so does professional identity. For a new scholar like Jessica, this uncertainty in the object translates to professional frustration and uncertainty. To see beyond the signal is to see the exoplanet as multiple and recognize the fragility of one's object of study; it is to expose worlds as things that are made. When visualizations alone fail to conjure a robust world, language games are played to aid in the practice of place-making.

Seeing through Language

Scientific inscriptions are always accompanied by verbal or written text.[22] As Seager mentors students like Devjit, Madhu, and Jessica in how to interpret novel inscriptions, she simultaneously provides a new vocabulary for them to use to speak about what they see. Much of this language is technical, but there is a subset of more colloquial terms, discussed in this section, that do more than just describe the data. They elevate the data to a level at which the detected planets take on elements of place. Anthropologist Kathleen Stewart (1996) has highlighted the importance of language in the place-making practices of Appalachian communities. She considers the importance of "signs," which Stewart broadly defines in order to understand the interaction between spoken, written, and social activities. Her interest is in the slipperiness of signs, "how, in the gaping sociality of signs, even the process of naming things does not so much fix a sign to a referent but marks the space of a gap between a *real name* and *what we call it*" (141). Appalachian names are signs with fluid, changeable meanings. Stewart describes signs not as explanatory but as "a way of reading likenesses" and "a gap in which strange associations remain possible" (146).

Despite what astronomers desire, there is a very large gap between the light curve and truly understanding an exoplanet as a world. Across this gap, astronomers weave "strange associations" as they make these alien planets familiar places. The naming schemes they use draw on analogy and metaphor, both in their written and spoken signs. Stefan Helmreich (2000), in his ethnography of Artificial Life computer scientists, tracks the use of metaphor in world-making. He observes, following Lakoff, that scientists, like all people, are creatures of language and as such are inescapably creatures of metaphor. Helmreich shows how scientists in conceptualizing computer simulations as self-contained "worlds" draw from the language of the living world and use the images that language evokes to describe and make sense of simulations. For exoplanet astronomers, language brings worlds into being. Language and metaphor mimic the work of visualization to make visible the invisible and create new realities (Tuan 1991; Lakoff and Johnson 1980). When not enough observational data exists to transform planets into worlds, exoplanet astronomers shape these planets through language. Linguistic signs help astronomers not only to "read likenesses" but also to offer the community a language with which to speak of and describe exoplanets as worlds and as places. In connecting planets to familiar places, these strategies alleviate some of the uncertainty and multiplicity introduced in modeling.

Scientific articles are riddled with terms such as "hot Neptunes," "eccentric Jupiters," and "super-Earths." When Jessica and Seager began writing their article on GJ 1214b, they frequently relied on the language of "Earths," "Neptunes," and "ocean worlds" to think through the possible compositions of the exoplanet. The associations provided touchstones, making the planet familiar as well as hinting of questions associated with such characterizations. They were at times uneasy with their use of analogy but had a hard time getting away from it. In an early draft of the article, they articulated that one interesting aspect of GJ 1214b was that there were "no solar system analogs" and "we should not carry over our own biases from solar system planets." Yet in the very next paragraph they could not escape relying on that language and describing how the planet might have formed "like Neptune." In figuring out how to describe the water planet scenario, Seager thought aloud during one meeting while deciding what to write: "Water planets are not like any in our solar system. Or water planets would be like bigger, hotter versions of Ganymede or Jupiter's icy moons." Each time they tried to get away from solar system analogies, it became apparent that analogy was the only way out of the semantic gap. In the end, solar system language served to organize the end member cases, and in the article they submitted they used the labels "Mini-Neptune," "Water Planet," and "Super-Earth." This language is not unusual in the literature. However, a referee suggested that they were overusing solar system analogies and the planet taxonomic classes they were relying on were not precisely defined. Consequently, in the final article, they renamed their cases "Gas-Ice-Rock Planet with Primordial Gas Envelope," "Ice-Rock Planet with Sublimated Vapor Envelope," and "Rocky Planet with Outgassed Atmosphere." Very much as Stewart describes, the original place-names used by Jessica and Seager did not offer fixity but did signal a gap between the known and unknown. However, in the end, familiarity was replaced by a more scientized scheme, despite the important role analogy had played in developing the project.

A conversation I had with an astronomer from the NASA Goddard Space Flight Center further demonstrates that familiar names are more readily understood as places. He described how he and another astronomer, at the start of the exoplanet boom, each independently wrote articles that came out within months of each other. Each theorized the existence of planets that were the same mass as Earth and at such a distance from their host stars that their surfaces were covered entirely in liquid. The astronomer I was speaking with had called these planets "Volatile-Rich Planets"; the other had called them "Ocean-planets." He reflected morosely that the term "ocean-planets" was widespread in the field and that article had garnered many citations, while his term had vanished from the literature. As our conversation continued, he realized why his article had failed where the other had succeeded: "ocean planet" was a familiar, imaginable world, while "volatile-rich planet" did not evoke an immediately recognizable kind of place. Something "volatile-rich" does not suggest a place, whereas you can imagine experiencing and being on an ocean-covered planet.

In addition to these naming conventions, the discourse of exoplanet astronomy draws heavily on language used to describe conditions on Earth. Even though many of the exoplanets discovered so far are gaseous giants, in discussing their properties astronomers struggle to make them seem Earth-like even if they are far from familiar. Speculating on weather and seasonal variation was a common rhetoric at meetings among the MIT exoplanet astronomers. In one meeting, we were discussing a theoretical planet where temperature is transferred between the atmosphere and surface. On one page of a handout about this planet was a graph of what the atmospheric temperature would be at different latitudes during "January." The speaker went on to say how this graph would be different in "March" and the other "spring" months. This object, which began at the beginning of the meeting as just a body orbiting a star, now could be imagined to have seasonal variation, and we had a better grasp of this alien world by understanding the difference between our spring and its spring. After we understood how volatile this planet's climate would be, an astronomer joked how alien people on the planet might react to the crazy weather—who would be the Al Gore of this exoplanet, she asked?

In one of Seager's research meetings, a graduate student described the atmospheric properties of a specific existing exoplanet he had been modeling. He mentioned that the planet is "tidally locked." This prompted another student to ask about the planet's "hot spot"—the part of the atmosphere that would be superheated by the star's direct and constant radiation. The student simply answered that the hot spot was "downwind" from where you would expect. Phrases like "tidally locked" and "downwind" are curious because of the physicality they imply. Tidally locked, for example, is an analogy drawn from our own Earth-Moon system. The Earth and Moon exert a gravitational pull on one another. One effect of the gravitational pull of the Moon on Earth is the ocean's tides. Another effect is that the Moon orbits the Earth in such a manner that the same side always faces Earth. This configuration is termed "tidally locked." Exoplanets close to their host stars often end up in a similar arrangement. Whereas in the Earth-Moon system, "tidally locked" references both a dynamic configuration between two bodies and an oceanic effect, scientists describing exoplanets as tidally locked are referring only to the configuration between planet and star. These planets, orbiting so close to their stars, cannot sustain water on their surface, nor do their host stars have oceans with tides in the same sense as Earth. Instead of introducing another term to the planetary lexicon, "tidally locked" is maintained for its powerful visual resonance. To think of tides and winds on exoplanets, even if they do not exist, allows one to imagine what it might be like to stand on the surface of one. Using the phrase "tidally locked" cues up a vivid imagination the community shares about what the planet might be like, how it moves, and what it feels like to be on it.

Speculating on the weather of exoplanets is a way to imagine the surface conditions. As astronomers imagine being on the surface, they transform objects into places. But, even if we could travel to these exoplanets, once arrived would we be able to set foot on their surfaces? The confirmation of the first rocky planet, the first so-called super-Earth, was announced in October 2009 using data from the CoRoT space telescope. This planet, CoRoT-7b, orbits twenty-three times closer to its star than Mercury does to the Sun. We would not be able to take a walk across the rocky surface, as the heat from the star likely keeps the rock molten. However, in the article announcing the refined mass and density estimates, this point was made: "If one assumes that CoRoT-7b is representative of the 'super earth' population . . . the structure of these planets is likely to be quite different from Neptune structure, but rather a more rocky planet like the Earth" (Queloz et al. 2009, 316).

The extensive use of metaphor in describing exoplanets—this pattern of using familiar names when discussing the unfamiliar—is common practice in exploration of the terrestrial sort. Geographer Yi-Fu Tuan (1991) notes how European explorers encountering the unfamiliar Australian desert for the first time still used the language of mountains they carried with them, despite the lack of significant elevation. Tuan asks geographers to look beyond the economic and material grounds for place-making and focus instead on speech and language. "Speech is a component of the total force that transforms nature into a human place. [Speech can] make things formerly overlooked—and hence invisible and nonexistent—visible and real" (685). For exoplanet astronomy, this "force" comes from the metaphors employed by planetary scientists that link their objects of study with familiar worlds. Their speech, in a very literal way, brings a reality to unseen objects and helps them understand what kinds of places their planets are.

Conclusion:
Seeing Is Believing

Keith Basso (1996b), in describing the importance of places and place-names for the Western Apache, takes the reader on the same journey he experienced, in which he learned to see the landscape in a new way as part of his cultural indoctrination. Likewise, over the course of my time with the exoplanet astronomers, I learned to see a squiggle or a little dip as first a planet and then a world. For both Basso and myself, a new way of seeing was intimately connected to ways of understanding place and community. Further, for exoplanet astronomers who are themselves in the process of creating a visual language, it is the ability to conjure worlds that reinforces the community. Worlding, for actor and analyst, gives tangible qualities to abstract ideas. A collectively held, fostered, and taught planetary imagination aids in training newcomers or visitors to the field how to see data as worlds; maps of Mars in a similar way convey a sense of place to a broader audience.

For exoplanet astronomers, learning to see (and developing this imagination) is a shared activity. The successful exoplanet astronomer is one who is able to discipline colleagues to see in a similar manner—a process akin to what Janet Vertesi (2015), a fellow ethnographer of planetary scientists, has called "drawing as." Not only did Seager pass these conventions on to her students, but when the MIT exoplanet group met, their manner of discussing papers reflected a community still reaching consensus over conventions of seeing. The ultimate goal is for a visualization to be seen as a planet that evokes a world: a "hot Jupiter," an "ocean planet," a "super-Earth." Language and visualizations work together to suggest not only that a planet exists but, more important, what the experience of visiting this planet would be like. All the while an imagination fuels this understanding of the planetary whole as a world.

"Is this real?" "Do you believe this?" A paper was being passed around on a Monday afternoon in June at an exoplanet meeting at MIT. This informal meeting was a weekly occurrence when professors, postdocs, visitors, and students met to discuss recent (and often controversial) discoveries in exoplanet science. We sat on couches in a circle on the ninth floor of the Green Building, enjoying the view of clouds rolling over the Charles River.

The eight of us present, slightly fewer than usual because spring term had ended, were discussing a paper announcing the first planet discovered by the astrometry detection method.[23] To assess the paper, we went through it figure by figure. With each figure, Seager asked whether it was "believable." We paused at the third figure: a scattering of points, with error bars depicting the radial velocity measurements. Seager said that we are "supposed to see a sine curve" in this graph (indicating the presence of a planet). One of the graduate students laughed, saying that he could also see a straight line because there were so few points and the error bars were so extended. In the fifth figure, the authors made the sinusoidal shape more convincing by fitting the data to a curve. The skepticism of the group, on viewing this figure, did not budge, and someone imagined other reasons for this sine curve: though it was supposed to represent a world, he suggested that maybe it was an anomaly in the optics. Finally, we came to the figure that was the crux of the paper's argument. This was not a typical representation in exoplanet astronomy but a display of the predicted Keplerian orbit for the exoplanet based on the ascension and declination measurements (locations in the equatorial coordinate system) of the host star for the eleven observations. The theorists, unfamiliar with this kind of representation, looked to graduate student Rachael, the most experienced observer, to interpret this figure. "I don't believe that orbit," she declared, but compared to the other figures, this plot was the authors' most convincing visual argument that a planet existed. But all it actually told us was that the star was in two positions, not that there was a planet orbiting it. Again, the figure failed to conjure a world. Our discussion of this paper concluded by pointing out a weakness of the paper overall: there were too many figures, and the authors should have done a better job in selecting what the reader needed to see. The visual argument did not come together.[24]

The language of "seeing" and "believing" was predominant in this discussion. Whether or not the authors can get their readers to "see" the data the same way they do is crucial for community acceptance. Planets must pop out of the graphs. The persuasive power of visualizations in science is well established, and this case makes such power quite stark. Lynch and Edgerton elevate representations from the status of "by-products of verbal 'ideas' or experimental logic" (1988, 186) to equal footing with the surrounding text in scientific and popular articles. My account illustrates how astronomers repeatedly focus on the images often without even reading the text. This suggests that in exoplanet astronomy the pictorial way of arguing is primary to the textual or verbal.

In each example I have provided—from Devjit and Seth's work with messy CoRoT data to Jessica and Madhu's complex models—translating these graphs into visual objects was not an easy or natural task, despite claims of "seeing the planet" in the data. Seager tried to explain: "A picture doesn't jump in my mind, it's really much more complicated. I've never really articulated this, so it's very challenging. . . . I think of the processes that are happening in, I don't know if it's a visual way, but in a very deep way inside my brain, so I don't know if I can communicate it." She tried to pick out a couple of diagrams (from her dissertation and from a presentation) to explain what she imagines when she thinks about a specific planet, but she couldn't quite convey the experience.

This gets at a central irony in exoplanet astronomy: despite the emphasis on "seeing," exoplanets are, for the most part, unseen. Unlike scientific objects that can be captured by photography or other means (Rudwick 1976; Latour 1987; Lynch 1991b; Canales 2010), in exoplanet astronomy no image of the object itself exists. Astronomers have thus crafted many different representations—from light curves to radial velocity graphs to visualized statistics—to stand in for planets. Exoplanet astronomy as a new visual culture is one with many layered and new ways of seeing. Seager elegantly described this way of seeing as understanding "data as art":

> It's sort of the way you would go and look at art. . . . You're looking at [the light curve] on many different levels, right? One is aesthetic appreciation, like "oh this is so beautiful I just want to swoon." Or it's so shocking or it's just vibrant. And then I'm looking for patterns. Like in [a painting], the artist often had something else in mind. Something deeper. So we have to look a level deeper. A level deeper is now not that simple appreciation, but it's what's really going on here. And it gets more technical. But all this happens almost instantaneously in a very mixed-up way. . . . When you say "what does a planet look like" this is what a planet looks like: a transit light curve.

However, the exoplanet astronomer does not stop wondering what an exoplanet is like once the light curve is produced. In working to understand their scientific objects, astronomers leverage abstracted graphs, mental pictures, and linguistic analogies. To make these distant objects tractable, astronomers craft planets and, ultimately, places. Astronomers ground their professional identities by making what they study seem less like ephemeral, distant objects and more like intimate, recognizable worlds.

One kind of exoplanetary world that continues to elude astronomers is a world that does not seem exotic or strange but eerily familiar. The discovery of a planet like our own is much anticipated across the community; following that quest brought me to yet another other-worldly location.

CHAPTER FOUR

INHABITING

OTHER EARTHS

I am standing on the "edge of the abyss," or *tololo*, in Aymara, the language indigenous to the part of Chile where I am. To the west stretch the desert mountains of the Andes, a natural border between where I stand in central Chile and neighboring Argentina. I am careful with my footing as I walk along the mountain's edge, kicking up dust and startling small songbirds. To the east, up a small rise on the 2,200-meter peak of Cerro Tololo, are the telescopes of the Cerro Tololo Inter-American Observatory (CTIO). The silver dome of the four-meter telescope glints in the sunlight. At night I find myself standing at the edge of what some might think of as a celestial abyss. Looking up at the stars, I place my hand on the wall housing the 1.5-meter telescope to make sure I do not fall over as I get lost in a view of the night sky as I have never seen it before. Earlier in the night a brilliant Moon outshone the clusters of stars and galaxies that are now observable with the naked eye. At one or two in the morning, as I step out of the observatory for the first time since the Moon has set, I gaze dumfounded at the sky, brought out of my silent reverence only by the creak of the neighboring telescope as its dome rotates to find a new stellar target.

I confess that this sort of romantic rumination was not typical of my experience, nor of the experience of other astronomers at CTIO. Moments before I found myself gazing up at the glorious night sky, I had been inside the observatory's control room, where I had been sitting for several hours. Perched in a dimly lit, windowless room with a linoleum floor and a drop ceiling, in the company of four other people and three times as many computer monitors, I had been struggling to keep my eyes open. My trip outside was practical—meant to wake myself up with the fresh, mountain air rather than invite a starlight-induced reverie. Yet it was this inspiring view of the night sky, rather than the computer work that occurred inside the control room, that seemed more of a piece with the grandiose goal the computations were seeking to achieve. The scientist I accompanied to CTIO, Debra Fischer, is one of several exoplanet astronomers scouring the night sky in search of a planet like our own orbiting a star like the Sun. Such a planet must orbit such a star at such a distance that liquid water, that all-important life-sustaining matter, can exist on its surface. This search for a "habitable planet" is a search for a very specific kind of planetary place. Such an exoplanet captures many aspirations of the planetary imagination that informs the scientific work of previous chapters.

The narratives spun at MDRS introduced how the planetary imagination is a product of past, present, and future conceptions of what it would be like to live on another world. The Mapmakers embed key features of this imagination into widely circulating maps of Mars, bringing a Mars that is a destination and a place of exploration to a broader public. As scientists begin to incorporate exoplanets into this same constellation of narratives and imaginings, the exoplanet community develops new visual and linguistic semiotics to facilitate world-making. At each of these sites, scientists made places by implicitly and explicitly deploying a sense of standing on or walking across a planetary surface. In this chapter I will more deeply explore this aspect of place-making by considering the ways "habitation" matters for astronomers. There are two senses of inhabiting that I will track here. The first concerns how modes of inhabiting are changing for astronomical practice. Earth-based observatories are no longer the preeminent epistemological place for astronomy. As space-based telescopes are increasingly used in exoplanet astronomy, practitioners struggle to articulate why inhabiting distant, mountaintop observatories is still important. A desire to inhabit observatories is at odds with technical necessity. Second, habitability is being constructed by these same astronomers as a scientific pursuit. A "habitable planet" is not a natural thing simply existing out there in the universe, waiting to be discovered; "habitability" must be imagined, defined, and made important. For the exoplanet community, "habitability," a rather unglamorous word, has become shorthand for what astronomers consider the greatest discovery their field, and possibly humanity, can make.

The story I opened this book with, of a young child pointing to a star around which a habitable planet is known to exist, illustrates how finding a habitable planet has come to be synonymous with a changing understanding of the universe and, moreover, a changing way in which humans might *be* in the universe. This narrative was first told by planetary scientist Jonathan Lunine during a testimony before the President's Commission on Moon, Mars, and Beyond, the name for the Bush-era space policy, in April 2004. Then, when only 120 exoplanets were known to exist, Lunine ended his testimony with two possible futures, both ten years off in 2014. In both, a scientist sits around a campfire with her children.[1] The first possible future is one in which NASA did not invest in the development of a dedicated mission to find a habitable planet.[2] The scientist mother, then, cannot definitively tell her children whether a planet like ours exists elsewhere in the universe. In the second future, NASA did develop and fly advanced detection satellites:

> The scientist walks their kids away from the campfire out into an open field and points to a certain set of constellations in the sky, and she points to two stars in particular and says, "Do you see these two stars? Each of them we know has an Earth orbiting around it, much like our own Earth orbits our sun. We know that there is air and there are clouds around that particular planet, the one around that star, and so there are plans to look more closely at it to see if there are signs of life." And then she concludes, "Maybe some day when your children's children's children are alive, they will go to that distant world to touch its soil and meet whoever or whatever is there" (Lunine 2004).

Here the aspiration of the exoplanet community is plainly told: one day, humans might inhabit another world.

The habitable planet is a specific kind of planetary place that today's astronomers are producing. "Habitable" and "inhabitation" are characteristics related to dwelling in a place. Philosopher Edward Casey presents the home as the archetype of human habitation, but he also suggests a more expansive definition. Inhabitation "is at home in all the places, actual or virtual, in which imagining and remembering flourish in felicitous space" (1998, 295). Even as the practice of astronomers is less defined by inhabiting a particular place, they direct their imaginings and rememberings of habitation toward defining and refining the particular, yet to be discovered, place known as a habitable planet. By chapter's end I will suggest that the search for a habitable planet in fact returns us to thinking anew about Earth as humanity's home.

Astronomers searching for other Earths do so on mountaintops or using space telescopes. Project teams range from a few researchers to many dozen. I went to Chile for a week in March 2010 to understand a modest project lead by Yale astronomer Debra Fischer. Fischer is looking at our closest two stars, hoping that one hosts an Earth-like planet. During the summer prior to my Chile trip, I attended a science team meeting held in a conference room at NASA Ames. This meeting of approximately forty astronomers was convened to discuss the first data received from NASA's Kepler satellite. Kepler is a space telescope launched in 2009 that continuously surveys 100,000 stars in search of an Earth-like planet.[3] In both ethnographic experiences, I witnessed the utilitarian ways astronomers discussed data that might contain the "holy grail" of exoplanet astronomy.

The different scales and locations of these projects suggest that scientists inhabit contemporary practices of "observing" in different ways.[4] Whereas astronomical observing used to require the inhabitation of an observatory, today these activities are rapidly being decoupled. The need to inhabit a physical space is declining just as the desire to detect a habitable planet is on the rise.

Location has always been epistemologically important to astronomy. The observatory used to be the undisputed authoritative site of astronomical practice.[5] Historians and early astronomers wrote observatories into a narrative of seclusion, describing a near religious communion for scientists with the stars, a feeling I attempt to evoke in the first paragraphs of this chapter. Percival Lowell wrote at the start of the twentieth century of the scientific purity afforded by newly established mountaintop observatories: "[The astronomer] must abandon cities and forego plains. Only in places raised above and aloof from men can he profitably pursue his search, places where nature never meant him to dwell. . . . Withdrawn from contact with his kind, he is by that much raised above human prejudice and limitation" (quoted in Lane 2009, 137). Drawing attention to this genre of romancing the mountaintop, historian of science Simon Schaffer has pointed out that histories of astronomy are often written as histories of confinement. Schaffer seeks to shatter this illusion of astronomy as a secluded science by offering instances of the social and cultural influences of observatories on the greater world in the imperial (not to be mistaken for empyreal) context (Schaffer 2010a, 2010b).

Today's places of observational astronomy include not only observatories (situated, as Schaffer suggest, in complex social and political networks) but also any place from which astronomers can access telescopic data. Today astronomers more often than not access data remotely. The observing eye—the technologies of observation—retreats ever further from the Earth's surface as observational practice becomes more withdrawn from earthly locales. To signal this changing configuration between observation and habitation I use the phrases *observing at, observing with,* and *observing from.* I begin with astronomers *observing at* CTIO, a grounded, mountaintop setting where they live at and fully inhabit the telescope facility. Kepler on the other hand orbits the Sun, trailing the Earth. When *observing with* such a satellite there is no observatory to inhabit, so the astronomer instead inhabits a sociotechnical network. The materiality shifts from buildings to Internet infrastructure, but both are still intensely social activities. My final stop is at a theoretical Archimedean point, an ideal point well beyond Earth's orbit. When *observing from* an Archimedean point, an astronomer inhabits a cognitive space, an imagined isolation neither grounded on Earth nor accessible via a network. This Archimedean point is imagined by theoreticians who ask what Earth would look like as an exoplanet. Here the observing eye is placed at such a great distance from Earth that when it looks back, Earth has dissolved into nothing more than a point. Can astronomers discern from this point that living beings inhabit the world? As the observing eye travels ever further away, it will be trained on exoplanets that seem more familiar—more Earth-like—to us than the alien worlds of the previous chapter.

Pursuing Habitability

Astronomers hope that habitability, in the sense of dwelling at and inhabiting places of science, leads to knowing. However, habitability, in the second sense of habitable planets, must also be established as a target that can be known. This is done first by establishing *where* habitability can occur, which has led astronomers to appropriate "habitable zone" as a cosmic descriptor. Terrestrial geographers and anthropologists of the late nineteenth and early twentieth centuries used "habitable zone" to describe climates suitable for human (and other fauna and flora) populations. The peaks of high mountains and large swaths of desert were beyond the "habitable zone" for these early scientists (see "Geographical Notes" 1886; Seligman 1917). The term trickled into planetary science by the mid-twentieth century, considering not the habitability of regions of the Earth but the potential habitability of regions elsewhere in the universe. Astronomer Su-Shu Huang was the first to designate a "habitable zone" on a stellar scale as the region around a star in which a planet would receive neither too little nor too much heat (Huang 1959).[6]

The "habitable zone" was very much a term of astrobiology (or exobiology, as it was then called) and was less prevalent among traditional planetary scientists. In the early 1990s, on the cusp of the detection of exoplanets, planetary scientists brought the concept of the habitable zone into a discussion focused more on planets than on the life they might host. Significantly, James Kasting, Daniel Whitmire, and Ray Reynolds (1993) published an article in *Icarus* articulating a definition for "habitable zone" based on the presence of liquid water. In the conclusion, the authors state the conditions under which the concept of "habitable zones around stars would assume a high level of significance." These conditions would be a successful result of either "the direct telescopic search for other planets" or the Search for ExtraTerrestrial Intelligence (SETI) (126). The first condition was met in 1995 with the detection of (the uninhabitable) 51 Peg b, and as the authors correctly predicted, "habitable zone" has become a defining term for the practice of exoplanet astronomers.

The habitable planet, a denizen of a habitable zone, is for astronomers a vivid world even though it is still largely an imagined one. The community writes about the significance of such a discovery with passion. In a 2008 report by the National Science Foundation's Exoplanet Task Force (a report in which task force member Lunine's story of mother and child once again appears), the authors suggested that discovering an Earth-like planet would complete the Copernican revolution. If such planets were detected, gazing at the stars would tempt us "with wild dreams of flight" and we would "refocus our energies to hasten the day when our descendants might dare to try to bridge the gulf between two inhabited worlds" (Lunine et al. 2008, 5). A similarly breathless statement was issued in NASA's "Exoplanet Community Report" the following year: "Astronomy has been an important preoccupation of humans for thousands of years, but the most profound questions—'are there other worlds and other beings?' reach back to the origin of *homo sapiens*" (Lawson, Traub, and Unwin 2009, 1). The authors of these statements suggest that the habitable planet has long existed as a place in the popular imagination.

For engineer Stephen Dole, habitable planets are places because they are destinations for humanity, as he explains in his 1964 *Habitable Planets for Man*. Writing post-Gagarin but pre-Armstrong (and with a hint of the burgeoning environmental movement, as he writes that Earth has already been exposed as "a tiny oasis in space"), he offers a quantified accounting of what is meant by "habitable planet." Defined simply, it is "an acceptable environment for human beings" (Dole 1964, 1).

For contemporary exoplanet astronomers, searching for habitable planets might be fueled by the grandiose aspirations of finding a destination or searching for life, but with respect to daily practice, it is the solution to a statistics problem. Inspiration might strike while gazing up at the night sky, but the real work happens in front of a computer, and discourse is dominated by methods of data processing and analysis. The Kepler telescope is focused on a sample of stars in one region of the sky that will be a statistical representation of the entire galaxy. At the science team meeting at NASA Ames, excitement grew not from romancing the idea of Earth-like planets but from revealing how clean the data were.

Fischer's project at CTIO, to find a habitable planet around Alpha Centauri A or B (our closest star system), relies on statistics and calculation in a different way from the Kepler project. Her research is designed to measure two stars constantly for several years. Once she amasses enough data, she will compile her observations and analyze them to detect a signal through the noise. At the observatory, the word "planet" was mentioned only a few times, and "habitable planet" was absent from conversation. Instead we discussed the instrumentation and sources of error in data streaming in from Alpha Centauri.

Yet Fischer designed this project because of the allure of finding a habitable planet around our closest stellar neighbor. The first time I saw Fischer speak in October 2009, she delivered an engaging talk to a group of scientists and social scientists gathered at the Radcliffe Institute for Advanced Study in Cambridge. She was practiced in describing her research, knowing how to interest a diverse audience with artistic renderings of the planets she has discovered and tales of the delight expressed by school-age children in the new planets. She calls her research on Alpha Centauri A and B "Project Longshot," because the odds of detecting a habitable planet in this star system are low. Even so, she is not alone in her pursuit. A mere sixty-five miles north of Cerro Tololo, the European Southern Observatory hosts the La Silla Observatory on a 2,400-meter peak. On a clear day, standing on Cerro Tololo and facing north, one can see the glint of the many silver observatory domes of the European institution. There, Michel Mayor (famed codiscoverer of the first exoplanet) is demanding that the High Accuracy Radial velocity Planet Searcher affixed to the 3.6-meter telescope spend part of its time focused on Alpha Centauri A and B. The close proximity of these two teams creates some tension. Fischer, during her talk at Radcliffe, joked that when she looks over at La Silla's peak she secretly hopes to see clouds obscuring their celestial view.[7]

Out of the statistics come wished-for discoveries. In the spring of 2014 Kepler announced its greatest discovery to date. The title of the article in *Science* read "An Earth-sized Planet in the Habitable Zone of a Cool Star." If this planet, Kepler-186f, were found to be rocky and have an atmosphere like Earth, it is likely that liquid water would exist on its surface. But such specifics about this planet will remain unknown, as it is located five hundred light-years away, well beyond the capabilities of today's remote sensing instruments. Several caveats keep the science team from claiming this as a truly Earth-like world. In the article's title alone there is no claim of habitability, only that the planet is within the habitable zone. Further, the host star is not like our Sun, and this "cool star" would likely create an environment quite different from our own. Even though Kepler-186f will remain only a potentially habitable planet, scientists have still imagined the qualities of its surface. NASA released a vintage-style travel poster inviting future explorers to the planet (fig. 4.1). Planetary visitors point across a white picket fence at a red forest, and the tagline reads "Kepler-186f: Where the grass is always *redder* on the other side." The red forest, the creator of this poster speculates, could be a result of a different photosynthesis process due to a cooler and redder host star.

4.1 Exoplanet travel poster: Kepler-186f. Image credit: NASA.

The fun of speculating about other habitable worlds occurs outside the day-to-day routine work of searching for them. The three modern-day "observatories" of this chapter—CTIO, Kepler, and an ideal Archimedean point—attend to these activities and gesture toward new presences associated with contemporary observing practices. Astronomers searching for habitable planets themselves *inhabit* each observatory in a different way. When observing *at* CTIO, astronomers inhabit the facility completely, as they live and work on the premises. Yet, as I will show ethnographically, the work done at the observatory is removed from the act of observing, as a physical divide remains between the instrument and astronomer. To observe *with* Kepler, meanwhile, a space telescope that is entirely uninhabitable, means to work solely through a sociotechnical network. Similar practices are associated with CTIO and Kepler, though Kepler is freed from the myth of seclusion signaled by the brick-and-mortar observatory in Chile. Finally, theoreticians seek to observe *from* an Archimedean point, which here represents an ideal observatory. Though it lacks materiality, astronomers cognitively inhabit this point in an effort to make progress toward understanding habitable planets even before they have been confirmed. I begin where the observers' gaze is most emplaced on Earth: I begin at the observatory.

Observing at the Cerro Tololo Inter-American Observatory

Debra Fischer enthusiastically welcomed me to join her on an observing run to CTIO. Along with two of her graduate students, Ethan and Jacob, she would be spending several nights observing and calibrating a new instrumentation setup in preparation for Project Longshot's main observing task. Though many other astronomers told me that observing was actually quite boring and they didn't know what I would learn by being there, I insisted that being bored on the top of a mountain in South America was itself interesting. I went to Chile in order to understand the place(s) of the observatory. Closely examining the literal landscape in which CTIO sits shows how the seclusion associated with observing at an observatory has to be constructed and maintained. Today's practices of *observing at* are not fully consumed by attending to the telescope. Ways of inhabiting the observatory are changing, yet astronomers still assert a need (or perhaps, more accurately, a desire) to be there. Describing CTIO's past and present serves to complicate the observatory's epistemological claim and at the same time demonstrates how alluring it still is to inhabit such a place.

I rendezvoused with Fischer and Jacob at baggage claim in the La Serena airport. I was flying in from San Francisco, taking a week away from my fieldwork at NASA Ames. Fischer and Jacob were arriving from New Haven. Fischer looked like a seasoned traveler in her grey slacks, a striped button-down shirt, and black polar fleece vest; all pieces were tailored yet loose-fitting. Jacob, in his California Academy of Science T-shirt and faded jeans, looked like a science grad student. After retrieving our bags, we found the car waiting to shuttle us to the CTIO off-mountain headquarters in La Serena.

The administrative facilities in La Serena, a two-hour drive from the tele-scope, were originally built in the early 1960s, a few years after the planning for CTIO began. The brainstorming for what eventually became CTIO began in 1958 after a Chilean astronomer approached astronomers at the Univer-sity of Chicago about establishing an observatory in Chile.[8] At that time there were only a handful of observatories in the Southern Hemisphere, none of them at a particularly high altitude. The Association of Universi-ties for Research in Astronomy (AURA), on behalf of the National Science Foundation, partnered with the University of Chile and by 1962 had, as a result of a multiyear survey led by Jürgen Stock, settled on the Cerro Tololo peak. After this decision was made, AURA began constructing the facilities at La Serena to house the unbuilt observatory's first administrator, science staff, and maintenance workers.

By my visit in 2010, the infrastructure had been much expanded. Cur-rently, a research complex houses the dozen or so resident astronomers as well as a few administrators and IT technicians. A lecture hall and library fork off from the main lobby. In addition to this building are several main-tenance buildings, houses, and apartments, as well as a "motel" where visit-ing astronomers stay. Our logistics coordinator directed Fischer, Jacob, and myself to this complex and gave us keys and room assignments. This nine-bedroom strip of rooms (each with a private bathroom) and commu-nal lounge was prefabricated in the United States and sent here in the late 1960s. I dropped my bags on an unadorned table and gazed longingly at the twin beds, but it was not yet time to rest. For astronomers visiting CTIO, La Serena is a liminal place—the observing trip has begun but one is not yet on the mountain. During our day in La Serena, we met and discussed progress on Fischer's project and instrument development. At night we stayed up late in preparation for the subsequent nights of observing.

Fischer is collaborating with Andrei Tokovinin, a resident astronomer who observes at CTIO on some nights and lives full-time on the La Serena complex. He is the instrument builder for the new spectrometer that Fischer is making to improve the performance of her telescope. As soon as we stored our bags in our rooms, Fischer, Jacob, her other student, Ethan (who had arrived several days earlier), and I went to meet with Tokovinin to discuss the status of the spectrometer.

Fischer designed Project Longshot to use the radial velocity, sometimes referred to as Doppler, detection method. In contrast to the transit method (favored by Seager's group and the Kepler mission), where one collects the fluctuating brightness of a star (photometry), radial velocity uses spectroscopy—the careful measurement of stellar wavelengths—so as to most accurately detect red and blue shifts. These shifts in the spectrum indicate movement of the star. A red shifted star (one in which light waves are elongated) is moving away from the observer; a blue shifted star (with compressed waves) is moving toward the observer. Slight changes in the shift, a wobble back and forth between red and blue, indicate the presence of an astronomical companion pulling on the host star.

Astronomers detect the subtle shifts in the spectrum caused by companion exoplanets, especially Earth-sized or smaller ones, in several ways. Fischer uses a detection technique she learned as a postdoc working for Geoff Marcy. Marcy and Paul Butler, when they first began trying to detect exoplanets, followed a technique pioneered by the Canadian astronomer Bruce Campbell. In order to measure the shift of the spectrum, a baseline is needed against which to measure. Right before the CCD collects an imprint of the starlight, the light passes through a cylinder of a gaseous element for spectral comparison. Marcy and Butler improved on Campbell's technique by using iodine as this element instead of the dangerous and unstable hydrogen fluoride (Lemonick 1998, 67–70).[9]

Fischer continues to use this technique. When she first began Project Longshot at CTIO, she was using a "vintage" spectrometer that, she said, had been "lying in the basement [of CTIO] just completely ignored and we dusted it off, polished it up, and sent a fiber to the 1.5-meter telescope." Even with the iodine cell, the resolution was not good enough for the precision she needed. But with what she called "NSF stimulus money" she won a grant to build a new instrument specifically for the telescope. The new spectrometer, which Tokovinin is fabricating, will be precise enough to detect an Earth-sized planet orbiting Alpha Centauri. Fischer, who had no experience with instrument design before this project, also works with two postdocs who were not in Chile for this observing run. These researchers were nonetheless able to participate in the "face-to-face" meeting with Tokovinin thanks to the video conferencing capability of Skype, which Fischer ran on her computer. As the meeting wrapped up, Tokovinin and Fischer both marveled at how much they had accomplished that day. This meeting had expedited what might have taken days to discuss via email. They concluded that there is something to be said for face-to-face collaboration. Even though astronomical work no longer requires astronomers to inhabit the observatory, they still benefit from being there. Habitation becomes less about instrument proximity and more about social interactions.

Our meetings ended around 4 p.m. Tokovinin was catching a 6:30 transport up the mountain and could not join us for dinner, but the four of us decided to go out in La Serena. As this was Jacob's and my first time here, we were excited to see what kind of place this was. As La Serena is located on the Pacific Coast, we decided to eat near the ocean where the resort hotels are. Elsewhere in the city are a university, a shopping mall with an English and Spanish movie theater, and a historic district where tourists (Chilean and others) go to buy craft goods and tour the nineteenth-century missionary churches. Our taxi drove us on the road adjacent to the beach. The hotels all displayed vacancy signs, indicating that even though the air was still warm tourist season had ended. We asked the driver to recommend a restaurant, and he dropped us off in front of an intentionally rustic-looking establishment. A sign hanging above the front door read "Gastronico." Once we entered and were seated, we noticed that the walls were adorned with astronomical pictures taken by the Hubble Space Telescope. Sitting beneath a spectacular picture of M51, the Whirlpool Galaxy, we realized that the name of this restaurant was a Spanish pun on astronomy (*astronomía*) and gastronomy (*gastronomía*). Our cab driver, it seemed, had a wry sense of humor in depositing the fare he picked up at the observatory headquarters at the astronomy-themed restaurant. Over pisco sours and empanadas we talked about places we had visited and would like to visit. Fischer was relaxed among her students, offering advice on relationships and mortgages. This easy night of conversation reflected her friendly, insightful, adventuresome, and supportive nature when discussing either science or life. A different cab took us back to CTIO's La Serena facilities, I said goodnight to my new friends, and though I knew I should stay up as late as possible in preparation for the next day's all-nighter, I could not resist falling into bed.

Ruta 41 and D-443

We had a noon transport up the mountain scheduled for the next day. The four of us loaded our bags into a minivan, and a driver was charged with getting us safely up to Cerro Tololo. Our van wound through the desert mountains as we climbed 2.2 kilometers to the peak. The scenery of this drive was startlingly similar to a drive I had made weeks earlier along the California coast on Highway 1, traveling from San Francisco to Los Angeles. Just as there, the road in Chile hugs the mountain, and if you swerve too far to the outside, only a thin guardrail is there to prevent your vehicle from tumbling over the cliff's edge. The only break from the dusty brown landscape was a vast, cerulean reservoir we encountered about a half hour outside La Serena. We marveled at the sight until the driver, noting our amazement, shook his head and sadly told us that to build this reservoir many small towns had been displaced. Their inhabitants had been forced to move up the hill and refashion their communities.

Though our time on the road was brief, the history of how this road came to be illustrates how ties between local custom and U.S. influence culminate in the greater mission of securing access to a clear night sky. Far from naturally secluded, the road has a history that emphasizes the work needed for CTIO to become a place associated with isolated and solitary observation. When Cerro Tololo was first selected as the site for the observatory, the only road to the peak was a mule trail.[10] Before a more sophisticated road was paved, it was a real journey to get to Tololo. It was a two-day trip from La Serena: three hours in a car and seven to ten hours walking or on a horse's back, with mules hauling equipment. The leader of the "seeing expedition" that selected Cerro Tololo and CTIO's first director, Jürgen Stock, described such travel as requiring "four wheels, four legs, and two legs."[11]

The ride I received took approximately an hour and a half and required only four wheels.

Stock, perhaps exhausted from hiking up and down Chilean peaks for two years, made road building a priority during the first years of CTIO construction. In late November 1962 AURA purchased the region surrounding Cerro Tololo, and road construction began in December. Due to the challenging terrain and to several changes of oversight, the road to the summit was not completed until September 10 the following year.

The president of AURA, Frank Edmondson, suggested they follow a U.S. tradition and hold a ceremony to inaugurate the opening of the road. The event, held in December 1963, featured speechmaking by local and national figures in Chile as well as by Americans, a ribbon cutting, and the sprinkling of holy water by the archbishop of La Serena. A second ceremony was held a year later in celebration of the completion of the observatory's cornerstone. This ceremony had for entertainment a choir of Chilean folksingers accompanied by women dancing Chilean national dances such as the *cueca*. The president of AURA fondly recalls a woman holding him firmly by the hand and teaching him the dance (Edmondson 1997, 162).

I am inspired to read the local, global, and universal twists in the road because of Peter Redfield's ethnographic work on the French space program in French Guiana (2000; 2002). His cross-reading of postcolonial studies with science studies seeks to articulate the terrestrial locality of outer space. Redfield writes about a strip of the only paved road in French Guiana, which passes in front of the space center where Ariane rockets are launched. He discusses the closing of this stretch of road and the debate it created between the space center (populated and run by French citizens) and the local authorities and activist groups. The debate centered on the spatiality of the road and its belonging to both the local (a local that is still fighting against colonial aftereffects) and the global/technological complex of the space program (which is pursuing the colonial enterprise beyond the globe).

Redfield explains that when the road closed, cars were circuitously diverted around the space facility. French Guianans interpreted this action as an estrangement: flow was disrupted, and traffic, formerly connected to the global enterprise of space exploration by this road, was reassigned to more local paths. The road leading to CTIO, in contrast, was meant to be an opening, symbolic not of estrangement but inclusion, connecting the local towns to the expanse of the universe. North American, colonial, and indigenous ceremonies commemorated the building of this road and the potential it had to make Chile a gateway to the cosmos.

However, the construction of the road, despite the ceremonial joining of different traditions, served to reaffirm the North Americanness of the Chilean telescope project. Before the road, Chileans were essential to the scouting and early building work led by Jürgen Stock. His progress reports detailed the local people and animals that supported his expedition. Not only did workers accompany him up the mountain to labor, on several occasions he describes small parties held on the peak for the few families that lived on the side of Tololo. One evening the Ramos family joined him on the summit with cake. They enjoyed a festive evening and set up a telescope so that "a view of the moon and Saturn concluded the pleasant day."[12] There was even the occasional uninvited (and unwelcome) guest who camped with Stock's crew on the summit. However, with the importation of machinery and the laying of concrete, local knowledge and animals were no longer needed to summit Tololo, and the observatory was marked as a place for North American astronomers.[13] Patrick McCray's history of observatory building in the twentieth century notes that CTIO, unlike the neighboring European Southern Observatory on La Silla, partnered directly with local universities and provided telescope time to Chilean astronomers. However, Chile's tumultuous 1970s decimated the university sciences, including astronomy, and there were few Chilean astronomers to take advantage of the telescope time promised at CTIO (McCray 2004, 239–40). The political turmoil that began in 1970 with Salvador Allende's controversial presidential election and continued with the military coup of 1973, out of which Augusto Pinochet claimed power, reaffirmed the impulse for European and North American astronomers to maintain CTIO as detached from its local place.

The director of CTIO who succeeded Stock in 1963, Victor Blanco, reflected on the rocky political situation of the 1970s. Though Blanco was shuttling to and from Santiago, managing local political currents, he insisted that CTIO stood apart. As he wrote in a brief memoir, "CTIO functioned normally during the time Salvador Allende remained in power, and not a single observing night was lost" (Blanco 2001, 12). Further, he recalled that only one or two observing nights were lost during the worst moments of the subsequent military coup. In recalling this tumultuous time, Blanco repeatedly asserted: "CTIO functioned normally" (14). A former director of AURA described CTIO: "It's a community that does nothing else but astronomy. It has its own culture and when you use it you temporarily join the culture" (quoted in McCray 2004, 238). These accounts insist that the observatory, just as Lowell claimed, was "raised above and aloof from men."

This separation, however, is a fiction. The road we drove up paved over connections between the observatory and local people that were necessary to CTIO's beginning. In the retrospective accounts of AURA's settling of Cerro Tololo, there are few mentions of the communities that were displaced[14]—and are still being displaced, as our van driver informed us on the trip up the mountain. Redfield, attending to the different things that happen in the same place, describes the complex, overlapping spatiality of the space program in French Guiana. Similarly, CTIO is not secluded on its mountaintop but is situated in a landscape with multiple histories and ties to the local, even if there are actions (intentional or not) that seek to exclude the local.

Cerro Tololo

Our journey up the road deposited us at the summit of Cerro Tololo in time for lunch. I dropped off my bags in a private room and walked down the breezeway that connected the dormitories to the cafeteria. The cafeteria was an airy room, made more so by the floor-to-ceiling windows at one end that looked over the Andean mountain range. These windows slid open so one could step out on the edge of the mountain. There were a dozen or so hefty wooden tables, some of which sat four while others were pushed together to accommodate six or seven. The Chilean staff of CTIO (administrators and maintenance workers, all men) always occupied the long table closest to the TV, which often broadcast a soccer game. I picked up a tray and utensils and went through the small buffet, selecting an assortment of salads, fruits, and desserts.

On the first day we were still day-shifted and lunch was lunch. On subsequent days this meal would be my breakfast (followed by a nap). Dinner became lunch, and the cafeteria would pack a bag of sandwiches and cookies and a thermos of tea or coffee for dinner at the telescope. Ethan, Jacob, and I always met for breakfast, but Fischer was absent, we assumed hard at work, during the day. We only saw her at the 6 p.m. "lunch" before heading up to the observatory.

On the day of our arrival, no afternoon nap was needed, and there were still several hours before the night of work could begin. Fischer had a Skype conference call scheduled, so Ethan, Jacob, and I explored the observatory. As we toured the facilities, we became familiar with the telescope that the next several nights of work would depend on. There was a stark opposition between the sublime landscape, with the Andes to one side and the ocean to the other, and the technological apparatus that was the telescope. The telescope was much different from the way I had imagined it and served to illustrate that even though astronomers place great value on inhabiting the observatory and being near the telescope, the instrument has become so complex that such cohabitation is a difficult task to achieve.

We left the cafeteria, setting our sights on the summit. We put on hats and sunglasses to shield us from the dangerously high UV exposure on the peak, and we hiked up from the dormitories toward the telescopes on the summit. As we walked up the dirt pedestrian path, we passed small clusters of telescopes that adorn the side of the mountain. The summit, which was leveled around the same time the original road was built, houses six telescopes. Fischer observes on the second largest, 1.5-meter telescope, which is adjacent to the silver domed 4-meter telescope (see fig. 4.2). Ethan, who had been to CTIO before, acted as tour guide for Jacob and me, who were encountering this environment for the first time. All three of us used cameras to capture both the equipment and the beauty of the natural surroundings.

The grandeur of the observatories, coupled with the minimalist geometric white markings against the dusty ground (reflective guides between telescopes used in the pitch black of night), evokes a technological sublime amid the natural sublime of the view of the mountains from the summit of one of the highest peaks. The awesomeness of the natural sublime of Kant and Burke gives way to, according to Kant, a realization of humanity's ability to dominate nature. In Kantian terms, the observatory is a resolution of the natural sublime, a material domination of nature leading to a rational domination of knowing the universe. It is also a manifestation of the technological sublime, as described by Leo Marx (1964). The observatory is perhaps even the quintessential South American technological sublime. Marx quotes a passage from 1844 that describes the sublimity of new

4.2 The technological sublime: Debra Fischer on the catwalk of the 4-meter telescope at the Cerro Tololo Inter-American Observatory. The 1.5-meter telescope is to the right. Photo by the author.

forms of transportation: "Steam is annihilating space.... Caravans of voyagers are now winding as it were, on the wings of the wind, round the *habitable globe*" (quoted in Marx 1964, 196; emphasis mine). With the railroad and the steamship, the globe began to shrink as humans traversed more of its area. The planet as a whole, beyond the places where people resided, became habitable.

More than a century and a half later, it is an iodine filter nestled in an observatory that is annihilating space and helping Fischer to find another habitable globe. Then and now, people viewing the globe and universe through the frame of the technological sublime experience a shrinking of space and an expansion of the reach of habitability. While human spaceflight extended the perceived habitable environment of Earth (Olson 2010), the search for other Earths extends the imagination of habitability even further into outer space. Jacob and I stopped to enjoy our first encounter with the technological and natural sublime, while Ethan urged us on to "our" telescope.

As we entered the white housing of the 1.5-meter telescope, the technological sublime was replaced by what might be called a technological mundane. The mountain vista was no longer present as we drilled deeper and deeper into the workings of the telescope and spectrometer. Ethan was eager to show us the instrumentation of our telescope. We dropped our bags in the control room and began our tour of the machine in the dome where starlight first enters the instrument. This was not the encased, streamlined telescope that the "gentleman" astronomers of centuries past peered into. The guts of this telescope, in the form of trusses and wires and parts I could not identify, poured out of the assembly. I would not have recognized it as a telescope. But the telescope did not end in the dome. We followed a fiber optic cable that connected the two parts of the telescope together. On one end was the assembly in the dome, and on the other end, encased in a room separated from the dome by several doors, was an equally complicated apparatus. This was Fischer's spectrometer (see figures 4.3a and b). In this cramped room, Ethan pointed out its parts: the iodine cell, collimator, echelle grating, slit, prism, camera optics, and CCD. The sense of triumph over nature was replaced by knowledge of how much meticulous work and testing went into this contraption. The technological mundane dominated the heavenly sublime.

Because the workings of the telescope were separated in space—half in the observatory dome and half in the instrument room—it was impossible to take in the whole system at once. It is also so finely calibrated that there were few parts we could move around. The experience of being at the observatory, as far as I could tell, had little to do with interacting with the instrument.[15] Yet this was the justification people often gave me for going to observatories. At the meeting in La Serena the previous day, Fischer had commented that once the new spectrometer is running and once data of a higher quality are being collected, she will have a team continuously at CTIO. She paused, then said that on second thought, the team could just as easily be at Yale. But, she went on, "there is something to be said for being here." The others at the meeting nodded in agreement, and Tokovinin recalled an "extremely rewarding night spent at Magellan [another Chilean observatory], seeing, touching the instrument." Tokovinin took on a hushed tone when recalling that night. He cited practical reasons for needing to be at the telescope; that, as an instrument builder, seeing the telescope firsthand allows him to better understand its outputs. And cer-

4.3 The technological mundane. (a, top) The telescope optics in the dome; (b, bottom) the spectrometer housed in a nearby room. Photo by the author.

tainly there are still practical and, as I will discuss, compelling social reasons for observing at a telescope. However, the allure of the mountaintop observatory is still connected to the sublime. On the first night, inspired by photos Ethan had captured of the Small Magellanic Cloud—a galaxy only observable in the Southern Hemisphere—Jacob went out with his digital SLR to take his own photographs of the night sky. Astronomers make their careers out of studying the mysteries of the infinitude that stretches out above and beyond Earth, but this is done by staring at computer screens filled with data and graphs. Being at the observatory affords one of the few chances to remember and reconnect with the awesomeness of a dark sky.

Yet during most of the time we spent observing, we were isolated from both the telescope and the glorious sky. Instead we worked almost exclusively from the control room.

The 1.5-Meter Control Room

The control room had low ceilings, had no windows, and was not much larger than a small classroom. Two tables were pushed together in the center of one of the walls, around which three people could comfortably work. Also along this wall were a printer, a minifridge, and a coffee station and microwave. On the other side of the room an L-shaped desk ran the length of two of the walls. On the desk were eight monitors, and above the monitors was a shelf. This was adorned with control boxes, a stereo, some journals, and other miscellaneous instruments. The room was lit with dim, yellow fixtures, inviting the feeling of nighttime even without windows. On one of the walls hung university pennants from Delaware, Stony Brook, Georgia State, Vanderbilt, and Yale.

Through the computers, a telescope operator controls the telescope and collects data from the CCD. Our telescope operator, Javier, was there for our whole observing run. Even when a visiting astronomer is not physically present, a telescope operator is at the facility. Fischer told me that most telescope operators at CTIO are Chileans with some kind of technical degree. They know a bit of programming and easily learn the specifics of the telescopes they operate. Though Javier spent most of his downtime watching TV, and his English was not very good (and only Ethan spoke Spanish), on occasion he would ask about the research we were doing and respond with informed questions.

During our first night, inspired by the novelty of being at the observatory, Fischer (after getting a refresher course from Javier) showed Jacob and me the digital sequence for moving the telescope between Alpha Centauri A and B. The observations for this project simply involve pointing the telescope at A, taking an observation (ranging from several minutes to almost an hour), and then moving the telescope to B, taking a similar observation, and repeating this process for the rest of the night. In other words, it is tedious. As Fischer's method is to combine data over the entire three years of observation, no significant data is being produced by any individual observation.

Consequently, it did not take long for the excitement of being at the telescope to wear off. Fischer and Jacob relinquished control of the telescope and allowed Javier to reposition it as needed. Like other "invisible technicians" (Shapin 1989), telescope operators are responsible for the smooth operation of experiments or observing runs, but aside from a "thank you" to the observatory staff in published articles, their role remains distinct from the astronomers' scientific work. In an interview, an astronomer described his relationship to the telescope operators: they are "professional observers who are happy to observe for us and they do as good or better job than we do. We just give them instructions and they do it."

It is because many large observatories have a staff of telescope operators that astronomers decreasingly need, themselves, to travel to the observatories. The same astronomer admitted that for routine collections, he does not need to be at the observatory. "But then sometimes it's much better to be out there." He elaborated:

> There are often a lot of decisions that you need to make during the night about whether to observe one target rather than another given the weather conditions, given the information you just learned the previous night. . . . You might just notice something at the telescope that someone who wasn't as invested in the project wouldn't have noticed, some error you're making or some better way of doing things. . . . And another important reason to go is to get to know the people who work there and who maintain these instruments and who build these instruments. That can be really valuable because you learn things that aren't in the manuals and those can give you a lot of really helpful information.

As this astronomer suggests, there are many different reasons to be at, to inhabit, the observatory. The visiting astronomer, when present at the observatory, can redirect the routine of the telescope operator in a way that might better serve his or her research. At the same time, one accrues social capital that might be helpful when not physically present at the observatory.

Because Fischer's and her students' tasks were not time or place sensitive, much of our time spent inhabiting the control room of the 1.5-meter telescope was not devoted to the search for a habitable planet. Instead we all worked on our own projects. Ethan was busy writing a paper that he considered his first real paper on exoplanets. Jacob spent some time working on a problem set for his class at Yale. For my part, I worked on a conference talk I was delivering in a few weeks. Fischer shifted between working on various collaborative projects, instructing Jacob and Ethan in their research, and explaining the finer points of observing and instrumentation to me. At one point her daughter called on Skype and chatted with us all.

Observing at

What is the significance of "being there" when "there" is not significantly different from the "here" of one's office? For anthropologists, "being there" is the essence of ethnography. Clifford Geertz ascribes the potency of anthropology, for better or for worse, to the ethnographer's ability to convey to the reader that she or he has achieved an "offstage miracle" and actually "been there" (Geertz 1988, 5). When discussing the degree of immersion in a foreign culture, Kirsten Hastrup (1995) acknowledges that most anthropologists assume that the more embedded one is, the better one's findings.[16] She quotes Margaret Mead: "As the inclusion of the observer within the observed scene becomes more intense, the observation becomes unique" (146). Echoing this sentiment, an astronomy graduate student described his experience at the observatory (in the "observed scene") with excitement. He was collecting data for an exoplanet known to transit and told me: "it's much more satisfying to see it [at the observatory]. You're like, 'I hope it transits,' even though you know it is going to transit because you know what you're looking for, but it's still kind of fun. Like, 'oh my God, a transit!'"

In "being there," the anthropologist hopes to know more intimately the people and culture about which he or she writes. I was at CTIO specifically to build rapport with Debra Fischer and study the relationships she formed on the mountaintop. Astronomers observing at the telescope are also there for reasons of social cultivation, even if their stated scientific objective is to study the night sky. Nights are spent in the control room, chatting with colleagues and the telescope operator, collecting data the astronomer likely will not analyze until back at the university. An important reason for going to the observatory is closer to the anthropologist's own reason for going into the field: astronomers build social relationships while at the observatory.

Even as astronomers increasingly observe remotely, running observations from their home institution or from a site near, but not on, the mountain, they attempt to cultivate a feeling of being there. The summer before I went to CTIO, I participated in a remote observing run with a graduate student, Rachael, at MIT. I sat with her for two nights while she accessed NASA's Infrared Telescope Facility in Mauna Kea, Hawaii. At MIT, students and faculty remote-observe from a room in Building 56 (the Department of Earth, Atmospheric and Planetary Science headquarters) that is equipped with the necessary computers and video conferencing equipment. Rachael, who has a lot of experience observing both at observatories and remotely, masterfully orchestrated the data collection. There were three monitors of importance in the conference room. Two were computer monitors perched on a desk, and the third was a television screen on an AV stand. The television displayed a live feed of the telescope operating room in Hawaii. On top of the screen was a webcam that presumably projected our setup onto a television or computer in Hawaii. We only engaged with the TV at the beginning of the night when discussing the evening's plans with the telescope operator, who popped in and out of our view for the rest of the night.

We mostly focused on the two computer monitors on our desk that were used to control the focus and exposure times of the telescope. During a particularly long exposure I asked Rachael about the room's setup. Why, I wondered, was there a second desk stacked on our computer desk? This created a somewhat cramped feel. Rachael laughed and explained that another graduate student wanted to re-create the feeling of being at an observatory. Sometimes, if you are remote observing with a telescope halfway around the world, the time difference works out such that you are observing during daylight. To create a more observatory-esque ambiance, the remote observer can close the makeshift blackout curtain, turn off the lights, and turn on the lamp mounted beneath the top desk. The observer at MIT can now feel that he or she inhabits the control room on the mountain. In effect, mimicking the arrangement of the control room makes the observer feel connected to a remote observatory and, as Mead offers, enables the observation to become unique. What is not, cannot be, mimicked is the clear night sky and that indescribable (though not scientifically necessary) feeling of connection with the universe.

Anthropologist Götz Hoeppe, who has also conducted ethnographic work in Chilean mountaintop observatories, writes of an email exchange with one astronomer who offered a material reason why being at the observatory is more potent an experience than remote observing: "I certainly do get a strong feeling of awe when the CCD reads out [at the observatory]. It is not at all the same when you go home and see it on the computer" (quoted in Hoeppe 2012, 1147). This astronomer explains that this strong feeling comes from her awareness that the photons coming from distant corners of the universe are not only falling on the CCD but on her body as well. As Hoeppe summarizes it, "the distant universe is, at the observatory, a part of the tangible world to which astronomers can expose their detectors and their bodies" (1147). Inhabiting, then, merges with observing in a way it cannot in remote practices.

Both Hoeppe and I note that being "fully there" at the observatory has little to do with the nuts and bolts of scientific work. I was much more involved in collecting data when I remote observed with Rachael than when I was at CTIO. Inhabiting the observatory has to do with creating the social ties that enhance scientific work: between teacher and student, astronomer and telescope operator, and astronomers from different universities. One goes to the (seemingly isolated) observatory to build connections. New technologies of observation have complicated this relationship between habitation and observation. The observatory is not the only place where astronomy can be practiced. Yet the legacy of the observatory makes astronomers hesitant to vacate it, and thus new uses of these spaces are now apparent. Inhabiting the observatory is shifting from an epistemological necessity to a social benefit.

With space-based telescopes, however, there is never an illusion that one can go to the observatory. This frees the astronomer from a need to observe or mimic observing at a specific facility. With Kepler, there is no control room to replicate. Observing with Kepler, the next stop for the observing eye that guides this chapter, does away with a brick-and-mortar research facility but further emphasizes the importance of inhabiting a network of people and data.

Observing with Kepler

At CTIO we rarely spoke of the larger project of finding a habitable planet. Perhaps, I thought, this was because we were performing preliminary data collection and mostly focused on instrument development. The promise of this project would not be borne out for several more years. My conversations with scientists working on the Kepler satellite also occurred at the beginning of the mission, but with so many stellar targets the analysis of these data had already yielded exoplanet discoveries. I attended a meeting of Kepler-affiliated scientists, the Science Working Group meeting, in July 2009 hoping to hear discussions about the importance of detecting an Earth-like planet. How, I wondered, did Kepler astronomers discuss these potential planets? In text on the Kepler website and in the news media, Kepler's goal of finding a habitable planet is portrayed as being of great significance for humanity. *National Geographic* ran an article about Kepler and other searches for Earth-like planets in which the author wrote that such planets "hold the promise of expanding not only the scope of human knowledge but also the richness of the human imagination" (Ferris 2009, 93). The meeting I attended did have an air of excitement, but this came not from speculating about habitable planets but from marveling over how "clean" the data were. The light curves presented were "beautiful" not because of planetary signatures but because of the precision achieved by Kepler's technological superiority to other space-based observatories.

Precision was also the goal at CTIO. The difference between Fischer's spectrometer and Kepler is that Fischer can make slight adjustments to the telescope based on data collection during the design phase. Once Kepler was launched, engineers could no longer adjust its optical performance. My distinction between *observing with* Kepler versus *observing at* CTIO encapsulates the physical disconnection between Kepler scientists and the Kepler satellite. One purpose astronomers say they have in traveling to observatories is to achieve an intimacy, or at least intensity, with their research. With Kepler, observational astronomy is removed from a specific location such that the sociotechnical network, which was also important at CTIO, becomes the astronomer's primary tie to the telescope. Before discussing the Science Working Group meeting in more detail and noting how data became a key node in the network that scientists fought to access, I will briefly narrate how Kepler became a space telescope in search of habitable planets.

Getting to an Earth-Trailing Heliocentric Orbit

The Kepler mission, named for a great astronomical thinker of the past, was ahead of its time. A decade before astronomers confirmed the existence of exoplanets, NASA scientist Bill Borucki was already determining the instrument specifications for the photometric detection of an Earth-sized planet. The promise of planets was thick in the air when he wrote: "detection of planets as small as Earth or Venus appears beyond the capability of ground-based systems, but might be possible when space-based platforms and extremely stable detectors become available" (Borucki and Summers 1984, 132). Astronomers in other subfields who were similarly stymied by limited technology convened in 1984 to discuss the limits and potential of high-precision photometry. Borucki gave a presentation on photometry's role in the search for other solar systems, specifically ones with Earth-sized planets (Borucki and Young 1984).

Over the next decade Borucki secured money from NASA to develop a series of photometric test beds, focusing his work on increasing instrument precision. In the early 1990s, NASA proposed a new funding model for exploring the solar system and beyond. Instead of NASA suggesting specific targets and piecing together missions from a variety of proposed spacecrafts, mission directors, and science projects, NASA introduced "Discovery-class missions." These missions were to be smaller in scope and would award funds to fully envisioned projects in which a principal investigator proposed a science target, mission architecture, and a team. Borucki thought his project well suited this new funding scheme. He gathered a team, designed a mission, and proposed to measure, as he put it in his proposal title, the "Frequency of Earth-Size Inner Planets" when the first call for Discovery-class mission proposals was announced in 1994. (This mission had already been rejected once before in 1992 by a NASA funding stream that preceded Discovery.) NASA declined the proposal, citing its similarity (in expense) to the Hubble Space Telescope. Borucki was not deterred. He continued organizing workshops and building instrument demonstrations to enhance the proposed mission. In 1996 the second call for Discovery-class missions was announced. Under the new name "Kepler," Borucki again proposed a space-based mission to detect Earth-like planets. Again the mission was rejected, and though Borucki worked to demonstrate feasibility, the Discovery program rejected Kepler for a third time in 1998. Kepler was not a mission for the twentieth century. Only after a complete technology demonstration and an abundance of exoplanets had been discovered was Kepler awarded a Discovery-class mission. In 2001 the idea for a space-based telescope that would survey thousands of stars in order to determine the frequency of Earth-sized planets, an idea hatched in 1984, was finally ready to be implemented with Borucki as the principal investigator (Borucki 2010).

Kepler was not always described as a search for "habitable" planets. Originally, Borucki referred only to "Earth-size planets." However, as "habitable zone" began to saturate the exoplanet literature in the mid-1990s, "Earth-size" was no longer a sufficiently interesting target. Three years after Kasting, Whitmire, and Reynolds (1993) offered a definition of "habitable planet" to the planetary science community, and a year after the discovery of the exoplanet orbiting 51 Pegasi, Borucki's planet hunting project embraced this metric. In an article describing his then unfunded mission, the strategy of detecting an Earth-like planet was linked to the concept of the habitable zone (Borucki et al. 1996). By the following year, an article discussing the (still unfunded) Kepler mission featured the term "habitable zone" in its title (Borucki et al. 1997). The banner of the official website for the (now funded) Kepler mission proudly displays the phrase "A Search for Habitable Planets."

Kepler was launched in March 2009. To limit the observational interference from Earth, Kepler does not orbit Earth but the Sun, in an orbit similar to Earth known as an Earth-trailing heliocentric orbit. Its photometric sensors point at a patch of sky in the Cygnus constellation (its "field of view"), taking measurements of 100,000 stars every thirty minutes. This generates a tremendous amount of data. But because one of the data sets might contain the signature of a habitable planet, the principal investigators and co-investigators tightly control access to the data. Consequently, the Science Working Group meeting was the first time that most members of the science team were seeing Kepler data.

Kepler was a topic of frequent discussion during most of my time with the MIT exoplanet community as astronomers eagerly anticipated the fulfillment of the mission's promise to find a habitable planet. Sara Seager is a participating scientist on the Kepler mission and corresponded with Borucki on my behalf to ensure my attendance at a Kepler Science Working Group Meeting in the summer of 2009. The meeting was two days long and was held at NASA Ames, where I would later return to work with the Mapmakers.

In a small lecture hall in Building 245 (where Carol Stoker has her office), roughly forty scientists affiliated with Kepler gathered early on a hot July morning. Most of the people present were affiliated with NASA Ames, SETI, or U.S. universities. A handful of Europeans were there, as one of the coinvestigators is from Aarhus University in Denmark and had successfully petitioned to use Kepler data for astroseismology research. The group was small enough that we all introduced ourselves at the start of the meeting. A neuroscientist and myself were the only two people who appeared "out of place" during these introductions. The walls of the auditorium were adorned with cosmic, terrestrial, and biological images, including glossy large-print photos of the Martian surface, a comet shooting through the Earth's sky, a volcano erupting, a close-up of a green tree frog. The scale from biota to planetary was fitting for a meeting of a mission designed to find a habitable planet.

This was the first gathering of affiliated scientists since the processing of the commissioning data. As such, there was anticipation in the air for what promised to be "beautiful data." To summarize the procedure of the two-day meeting, an astrophysicist deadpanned that like any NASA talk, this one would start by claiming success and then work back from there. Appropriately, Borucki welcomed us to what he said would no doubt be an exciting meeting, announced the flawless performance of Kepler thus far, and then, before diving into the planetary candidate light curves, worked back to the operational level of the Kepler mission. As such, the first talks of the meeting were delivered mostly by NASA employees and discussed activities at a management level. As the meeting progressed, "the professors" (as a university professor described his ilk to distinguish them from NASA scientists) increasingly assumed the podium and began replacing the flow charts on PowerPoints with equations and graphs. Operational-level discussion gave way to physics and science questions. There was no discussion about the likelihood of finding a habitable planet or even a revisiting of what "habitable" means. The term had long since stabilized and was here invoked simply as something that could be detected. All of the talks were about data calibration, data processing, the data pipeline, and who had access to what data when.

Being on a large science team like Kepler can be frustrating.[17] Often university astronomers work on projects like Fischer's, where one has complete control over data and can go to the observatory if warranted. Without the physical observatory, Kepler scientists' only connection to the project is through data. To inhabit Kepler (as one inhabits CTIO) requires access to the data. The principal investigators and coinvestigators kept data secretive during the first few months. The data were not being circulated, which meant physical presence at an authorized computer bank was necessary in order to do work (or even view the light curves). From NASA's point of view, this controlled against leaks and possible "scoops." I quickly realized that many of those attending the meeting, members of the science team, did not have permission to see the Kepler data. As with Mars data, NASA is contractually required to release publically Kepler data after a set number of days. Those present were quickly becoming anxious that the public would have access to the data before or at the same time that they did.

"Public" became a word of negotiation between the academics and NASA. When Borucki tried to assure the audience that there was a system in place by which data "goes to the public" (where this public included the scientists in the room), the academics countered that they should have lead time before the public release to examine the data. They attempted to distinguish themselves from NASA's use of "public" in order to gain a more advantageous position with respect to data. Speaking with a member of the science team after these discussions, I asked how the scientific public was different from the general public. The astrophysicist responded defensively that all members of the public have access to the data NASA releases and are, in fact, quite interested in this mission. Public, in this setting, was a blurred concept meaning either all interested consumers of (scientific) news, working astronomers, or working astronomers not including members of the Kepler science team. The conversations on public data consumption concluded with the formation of a committee to further address this problem and hopefully assuage the frustrations of the academic members of the noncore science team.

Without access to data, scientists did not feel like they were part of the mission. They could not inhabit Kepler's sociotechnical network because they had access to neither the observatory nor its products. Because it was a space telescope, they could not observe *at* Kepler, and without access to data, they were unable to even observe *with* Kepler.

Space telescopes shatter the façade of the myth that being in close proximity to the instrumentation is vital to current astronomical practice. With this forced remove between instrument and scientist, the act of observing is distilled to one's ability to navigate within (in the case of Kepler) a large and distributed scientific team. One advantage of the physical observatory, as I have mentioned, is the sociality fostered there between scientists. The team meeting still offers an opportunity to come together, suggesting that more than offering a place of practice, observing with Kepler offers a place of sociality. Instead of the place of work being at the observatory, it can be at any conference facility. However, the physical coming together is not in itself enough to fully stand in for the role of the observatory. Astronomers desire access to the data in order to feel that they are observing with Kepler. Inhabiting the distributed network of either ground- or space-based observatories is as much (if not more) about the abstract notion of access to data and the social networks that care for and interpret that data as it is about the concrete, physical location. As the observatory changes, so do attending forms of habitation. Inhabiting becomes ever more disconnected from a terrestrial place.

Observing from the Archimedean Point

The drama of the search for a habitable planet is still unresolved. Fischer and her team observe at CTIO, measuring the Doppler wobbles of Alpha Centauri A and B. The NASA team observes with Kepler, systematically going through their survey of 100,000 stars hoping to find a telling transit. Both teams inhabit networks and places on Earth to search for habitable planets. This is primarily a game for observational astronomers, but theoreticians have carved out a complementary project. Carl Sagan (1994) refigured the planetary imagination when he asked readers to ponder the Pale Blue Dot, the picture of Earth taken from so far away that it looks like a star.

Theoreticians have risen to this challenge, producing articles and books that treat our home planet Earth as an exoplanet. The Earth-as-exoplanet line of argument takes place neither on a mountaintop nor in a conference room but at an imagined extraterrestrial Archimedean point. To observe from an Archimedean point, theoreticians cognitively inhabit a place not of this world.

The Earth as a Distant Planet: A Rosetta Stone for the Search of Earth-like Worlds (Vázquez, Pallé, and Montañés-Rodríguez 2010) is a book in this genre of theorizing. The cover of the book features a spiral of Earths (a take on the famous "Blue Marble" image of the cloud-covered African continent taken during *Apollo 17*) against a black sky dotted with stars. This evocative image resonates with a *New Yorker* comic that first appeared in 1991 and exoplanet astronomers often use in their public talks. In this comic, a spaceship is flying through a dense field of Earths (this time, the North American side is the dominant face); the caption reads "Well, this mission answers at least one big question: Are there other planets like ours in the universe?" The joke points to an improbable scenario in which the discovery of an Earth-like planet will be truly a twin. Yet the iconography of multiple literal Earths remains popular in science literature, as this book's cover illustrates.

To study Earth as an exoplanet, astronomers take Earth data, either in the form of satellite images or "earthshine" (light from the Earth reflected off the night side of the Moon), and reduce these highly resolved data sets into a single point source to mimic our perception of exoplanets. Then they run various models to see if it would be possible to extract the complexity they just erased. *The Earth as a Distant Planet* provides several examples of this spatial reduction. Whereas beautiful satellite images of continents and oceans illustrate the first chapters of this book, representations of Earth in the chapter titled "The Pale Blue Dot" are light curves and spectra, shown in line drawings that are meant to resonate with the data that astronomers currently produce for exoplanets (as I described in chapter 3).

The research described in this book is by no means fringe and informs articles in top-ranked astrophysics journals. Sara Seager coauthored an article with the book authors to demonstrate that "the light scattered by the Earth to a hypothetical distant observer as a function of time contains sufficient information to accurately measure Earth's rotation period" (Pallé et al. 2008). Extending this line of thinking, astronomers have also sought to harness satellites beyond low Earth orbit in order to study Earth as an exoplanet. For example, in an article titled "Alien Maps of an Ocean-Bearing World" a group of university astronomers in conjunction with the science team of a NASA satellite designed for close encounters with comets used Earth data from this satellite to see if it was possible to distinguish oceans from landmasses (Cowan et al. 2009). The "Ocean-Bearing World" is in actuality Earth, and the "Alien Maps" are highly abstracted representations of Earth. As the satellite's observational data offer higher spatial resolution than exoplanet data, the first step was to integrate over each image, reducing them to single pixels. Then one takes the position of a naïve observer, assuming "no prior knowledge of the different surface types of the unresolved planet" (917). After examining the spectra, the team "discovered" that at times the planet appeared optically blue and at other times optically red, suggesting two kinds of surface types. The final step of the analysis was to construct an alien map of this ocean/land planet. Figure 4.4 appears in the article as a comparison between an actual map of Earth and the alien map reconstructed from spatially unresolved data.

The alien map suggests we are viewing something "other," but in actuality we are viewing a representation of ourselves. Representing the self as the other brings to mind the pendulum Michael Taussig (1993) provocatively swings in his book *Mimesis and Alterity*. Taussig asks, "What does such a compulsion to become Other imply for the sense of Self? Is it conceivable that a person could break boundaries like this, slipping into Otherness, trying it on for size? What sort of world would this be?" (33). This question launches Taussig into an analysis of a 1935 surrealist essay by Roger Caillois. Caillois diagnoses mimesis of the self as "being tempted by space," tempted by an unboundedness in which the self is but one point among many—one Earth among many. Caillois warningly writes that the mimed self "tries to look at himself from any point whatever in space. He feels himself becoming space. . . . He is similar, not similar to something, but just *similar*" (quoted in Taussig 1993, 34). The mimetic Earth—the portraying of the planetary self as the planetary other, the alien map—leads, for Caillois, to alienation.

But for scientific thinkers making the Earth alien leads to a deeper connection with our own planet. Decades before these astronomers produced alien maps of Earth in an effort to understand exoplanets, James Lovelock (1979) similarly engaged in an exercise of "othering" when he thought to search for signs of life on Earth as a model for how such a task might be

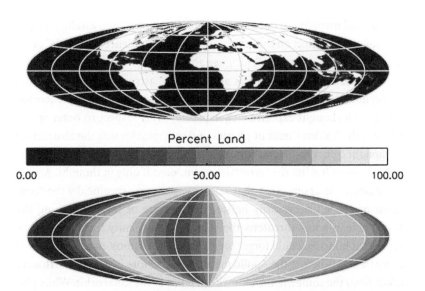

4.4 Alien map of an ocean-bearing world. (a, top) A cloudless representation of the Earth; (b, bottom) map reconstructed from reduced EPOXI data. Though (b) is not latitudinally resolved, it represents some ratio between land and ocean. Image credit: Cowan et al. 2009, fig. 10. © AAS, reproduced with permission.

accomplished on Mars. Taking the global view of terrestrial life led him to develop his popular Gaia hypothesis, which surmised that both organic and inorganic materials on Earth feed off of each other to create one large organism at the scale of the entire planet. Lovelock's attempt to treat the Earth as an alien world, though initially a strategy for understanding a different world, ultimately taught him more about our own planet.

Lovelock and the authors of "Alien Maps" could not observe the Earth as a globe from their position on its surface. As Taussig warningly describes, they are "spaced out." As the authors write in "Alien Maps," "These data reveal Earth as it would appear to observers on an extrasolar planet, and can only be obtained from a relatively distant vantage point" (Cowan et al. 2009, 916). This notion of a "distant vantage point" appears repeatedly in *The Earth as a Distant Planet*. Those authors talk of an "extraterrestrial observer," suggesting that the Earth-exoplanet connection must be made by first assuming a remote gaze from which to observe Earth and then returning to Earth and refocusing the gaze outward. They historicize the position of the observer and the viewing of Earth-as-planet in their first chapter. From on-the-ground mapping to hot air balloon photography, from satellite images to Apollo, they reflect on the potency of seeing "our planet as a whole" (Vázquez, Pallé, and Montañés-Rodríguez 2010, 12).[18]

When writing about Earth as an exoplanet, astronomers observe from an Archimedean point. Hannah Arendt (1963) wrote of her apprehension about such thought experiments, especially with respect to outer space. For her, the hidden threat of physics and astrophysics was the abstraction of thought from reason and common sense, which inevitably happens, she argued, when leaving the terrestrial realm, even if only in thought. Arendt questioned the assumption that scientific pursuits, specifically the "conquest of space," necessarily bolster the "stature of man." Pursuing the Archimedean point threatens the erasure of humanity from life, in the sense that humans would forget their *place* in the cosmos. Arendt expressed these thoughts before the Apollo missions and before images of Earth were taken from the somewhat Archimedean point of the lunar orbit. While philosophers showed concern over this disembodied view (Heidegger [1966], as another example, expressed fear when confronted with images of the Earth taken from the Moon), scientists enthusiastically anticipated such an extraterrestrial positioning. Vázquez and colleagues quote astronomer Fred Hoyle circa 1950: "Once a photograph of the Earth, taken from outside, is available. . . . once the sheer isolation of the Earth becomes plain to every man whatever his nationality or creed . . . a new idea as powerful as any in history will be let loose" (quoted in Vázquez, Pallé, and Rodríguez 2010, 11).

Four decades after the isolation of the Earth became "plain to every man," exoplanet astronomers are attempting to overcome this sense of isolation by searching for proof of other habitable planets. The Archimedean point is no longer afloat (as Arendt describes humanity floating, from Copernicus suspended in the Sun to Einstein's man drifting freely in space) but is finding ground on other Earths. For Arendt, the threat of the Archimedean point is that "once arrived there and having acquired this absolute power over his earthly habitat, [man] would need a new Archimedean point, and so ad infinitum. In other words, man can only get lost in the immensity of the universe" (1963, 278). However, the Archimedean point for exoplanet astronomy is not forever elusive. In searching for exoplanets with solid, Earth-like surfaces, astronomers seek an alien Archimedean point with a foothold on another planet. Contrary to Arendt's or Caillois's fears, habitability elsewhere does not forever condemn humanity to endless wandering but, as astronomers see it, offers a firm notion of Earth's place in the universe. It is from this vantage point, albeit one grounded on an imagined planet, that astronomers theorize about Earth as an exoplanet. *Observing from* the Archimedean point requires astronomers to cognitively inhabit this point, such that it makes sense to "look back" and observe Earth.

The Earth as a Distant Planet contains a chapter on how to detect whether or not life exists on such a habitable planet (eerily titled "Biosignatures and the Search for Life on Earth") and a few sections scattered throughout on the biology of habitability. But mostly the book is about Earth as a planet; Earth as a geologic and geographic entity. Likewise, the article "Alien Maps of an Ocean-Bearing World" mentions life signatures only in passing, and the "Alien" in the title refers again to a geographic, not biologic, other. In these moments, when astrobiology is put aside, another project shines through: the search for another planetary home. Making the connection between current practice and future practice, the authors of *The Earth as a Distant Planet* write: "In the near future, the application of sophisticated methods will give us the opportunity to detect and study planetary bodies similar to our Earth. The day we arrive at another Earth-like planet, we will already have in our possession detailed cartographic maps of it, taking advantage of the knowledge we have acquired in finding out about our own planet" (5).

Conclusion:
Home Worlds

In searching for habitable planets, astronomers confront the changing ways they can inhabit their places of science. No longer do they simply dwell at the observatory; they also inhabit increasingly distributed sociotechnical networks, learning to observe at, with, and from different physical and imaginative configurations. The dis-placing of practice from the defined site of the observatory is a trend mirrored in other sciences that are moving toward finding answers in "the cloud" and through the processing of Big Data. Inhabiting, in the sense of dwelling, is a fading mode of scientific, at least astronomical, practice. But, there is still a *desire* to be in a specific place. Going to the observatory will not soon fade from practice, but it will be driven more and more by a need to be in place and connect to the history of the profession and less and less by a scientific need. This desire for place is a component of the other sense of habitation developed in this chapter: the making of habitability as the definitive metric of an exoplanet that would have the greatest human significance. Both desires are about finding an antidote to a perceived sense of uprootedness. Most astronomers, like many citizens of the developed world, do not spend their lives in their hometowns or even their home countries. Further, the sentiment that you can't go home again is taking on new meaning as Earth itself, our home planet, seems very different, atmospherically and geologically, from the place it was fifty, forty, or even ten years ago. The search for an Earth-like planet is shot through with nostalgia—nostalgia in the precise sense of acute homesickness. Though much, perhaps impossible, work would be required to undertake a "journey home" (*nostos*, in Greek) to the Earth of the past that we think we recall, we can, perhaps, find an exoplanet that reminds us of an Earth more habitable than today's or the future's Earth. The search for a habitable planet is nostalgia for an Earth we have never known. It is a search for an idealized home.

The way home relates to human being has been theorized in different ways.[19] Martin Heidegger and Gaston Bachelard both write about the connection between home, the world, and the cosmos. Heidegger's phenomenology, particularly as expressed in *Being and Time* (1927) and "Building Dwelling Thinking" (1951), puts forth that dwelling is the primary mode of human being. In "Building Dwelling Thinking," Heidegger illustrates the power of dwelling with an example from the Black Forest in Germany, where he was born and to which he later returned. His home offers an example of an essential form of inhabiting and thus of being-in-the-world. In *The Poetics of Space*, Bachelard (1958) claims: "The house image would appear to have the topography of our intimate being" (xxxvi). The key to understanding the greater world, Bachelard insists, is to first consider the home. "[Our house] is our first universe, a real cosmos in every sense of the word" (4). The home image pervades one's life always and in every way. Ultimately, "all really inhabited space bears the essence of the notion of home" (5).

Immigration and flows of globalized movement have disrupted the twenty-first-century "notion of home." For Heidegger, this might signify a loss of essential modes of being. However, geographer Doreen Massey (1994a) sees this new geospatial configuration as an opportunity to recontextualize "home." She embraces the spatial upheaval as a way to think about place and home not as static and bounded but as products of social interactions, as parts of flows, not fixities. For Massey, you can't go home again because there is no singular idea of home.

As exoplanet astronomers write and talk about habitable planets, they are caught between Heidegger and Massey, between Home and multiple homes. A habitable planet is a promise of finding home, of finding a planet suitable for housing life as we know it. Perhaps astronomers will detect an Earth-like planet still in the early stages of nurturing life. This would be a glimpse of the foundation of our planetary house laid long ago. But in astronomical practice, the search for habitable planets is also entrenched in ideas of unboundedness, multiplicity, flows, and networks. Astronomers who are observing at CTIO and with Kepler inhabit networks that span the world and extend into space. They are seeking to "understand our place in the universe." But to do so it is sometimes necessary to break free from our home planet, as theorists do when observing from an Archimedean point and turning the gaze back on Earth. The planetary imagination, especially with regard to habitable planets, is never about a singular world but is about the potential for all planets to be worlds, our own Earth included.

CONCLUSION

NAVIGATING

THE INFINITE COSMOS

I began my fieldwork for this project in 2009. As I write this conclusion, more than five years later, much has changed in the field, and much has stayed the same. The MDRS is still in operation, and its visitors are still drawing connections between the Utah and Martian terrains. A crew member from the final mission of the 2013–2014 season, on encountering a murky river, has written, "So there it was, water on Mars and in Utah. We had done it. And for our purposes it's basically the same place" (Morgan-Dimmick 2014). Utah continues to be made into a place understood as Mars. On the Mars on which humans have yet to set foot, though, water and other signs of life are still being sought. The Mapmakers continue to serve up images and virtual globes that allow scientists and nonscientists to engage in remote exploration. Just as the rover *Curiosity* slowly inches toward its destination of Mount Sharp, the Mapmakers continue to stitch new high-resolution images into their quilted representation of a landscape, transforming it from the alien to the familiar. Both efforts are aimed at knowing another world in ever greater detail.

Exoplanet astronomy, a much newer discipline than Mars science, has moved at a faster clip. With so many known planets there are more data than ever before, and scientists work hard to analyze, scrutinize, and play with what can be discerned from dips in light. In 2009 the community felt they were on the brink of something great. The first rocky planets were being confirmed, and the newly launched Kepler space telescope promised to return data that contained evidence of planets close in size to Earth. Though Kepler malfunctioned in 2013 and is no longer precise enough to collect new data streaming down from Sun-like stars, the years of data already collected have indeed contained traces of planets that appear to be similar in size and perhaps even composition to Earth. Yet the greatest discovery—that of a true twin to Earth—remains elusive. The community is still perched on the same brink as 2009, as the question of what else is out there remains unanswered.

Finding and knowing other planets requires understanding large conglomerations of rocks and gas as worlds, as places. But why is place the framework being invoked to discuss these astronomical objects? Why does place matter? This is a difficult question to answer and one that has both an epistemological and an ontological answer. It has been my goal in this book to offer a compelling epistemological answer. Place, I argue, is not just a passive canvas on which action occurs but an active way of knowing worlds. Even when place is not self-evident, as perhaps with invisible exoplanets, it is nonetheless invoked and created in order to generate scientific knowledge. Place breathes meaning into alien worlds because it makes these worlds familiar and, moreover, familiar as something that is physically explorable. Places are exciting because we know how to know them; we all have experience learning what it means to be somewhere. Moreover, planetary science is not alone in leveraging the language of place as a scientific heuristic. Neuroscientists are "mapping" the brain, and nanotechnologists talk of "nanospace." In calling science the "endless frontier," Vannevar Bush (1945) offered a metaphor that spatialized scientific endeavors in order to make them conquerable. Thus, I am not arguing that the place-making performed by planetary scientists is unique to the field but that this book represents the first ethnographic attempt to detail and understand the scientific value of such activities.

What is unique to the field of planetary science is the ontological answer to the question of why planetary scientists have internalized place-making as a fundamental way of knowing planets. I have developed the idea of the planetary imagination to capture the continuity across my field sites of a desire to intimately know planets as worlds on which one can imagine being. But this imagination indicates more than just pervasive practices of worlding; it also speaks to a larger spatiality, one that extends beyond a single planet. I opened the book with the image of the next generation excitedly pointing to the sky, an image I revisited in chapter 4. This indexical gesture of connection requires knowing that something is there at which to point. I read this movement as an embodiment of a changing geography of the cosmos, a gesture that reflects the sentiment repeated again and again by exoplanet astronomers: that finding an Earth-like planet will change how humans view "our place" in the universe. Figuring Earth as one of many places that humans can be on, one of many places in the universe, creates an interplanetary network that planetary scientists have become comfortable inhabiting.

When I asked my interlocutors why this remapping of the cosmos was necessary or meaningful, conversations that had been nuanced and personal turned abstract and stilted. In one typical example, I was asking a graduate student about the relationship between exoplanet astronomy and astrobiology, and he quickly responded that "in the end, everyone wants to know if there's other life, so I think even with me [studying the shape of exoplanets] that kind of drives what I'm doing. In the end we want to find other Earths, we want to find something similar to us." When I asked him to explain this drive for finding other Earths or other life, he searched for an appropriate answer:

> Uh, I don't know, uh, relevance? Our own? I don't know, in some sense, uh, I don't know, the scale of things? I don't know, the basic question of how common we are? I'm not a very religious person, so for me it just seems likely and I don't want to feel anthrocentric [sic] or Earth centric in some sense, I want to see that validation that you know it's more than just us in some sense. I don't know, that's not very articulate. It's just interesting, childlike interest. I don't know. This is the same interest I had when I was really young so obviously it must be something important.

This statement mixes together different ways one might frame one's professional goals: from the quantifiable question of how large the universe is to the abstract notion of humanity's relevance in such an expansive universe. The way this graduate student reflected on his chosen path calls to mind a vocational calling, one that was, for the student, unavoidable precisely because its profound implications were "obvious."

By way of conclusion, I wish to reflect on the "obviousness" of the objective of planetary science's search for worlds and life like our own. What do understandings of what is "out there," the spatiality that looms larger than the planetary imaginations and what we also might call cosmologies, tell us about what it means to be in the universe? I will suggest that today's dominant astronomical cosmology is presented as a mode of connection, made possible through the transformation of planets into places such that it becomes a possible task to know our *own* place in the universe.

This generation is of course not the first to ask after the existence of other worlds. As many scholars have documented, this question was asked in antiquity and has rarely receded from the philosophical, theological, and scientific agenda (Dick 1984; Crowe 1986, 2008). My own interlocutors made frequent references to one of the earliest askers of this question. As I mentioned briefly in chapter 3, exoplanet astronomers regularly quote Epicurus's musings on the existence of infinite worlds. In his letter to Herodotus, which astronomers quote, Epicurus's statement about many worlds is made in the broader context of outlining his theory of the material world consisting of infinite "atoms." For Epicurus and other atomists, evidenced by the later writings of Lucretius, the existence of other worlds was a conclusion to be drawn from their theory of matter, not a sought-after answer to a specific question.

The atomists believed in an infinite, unbounded universe (one that thus must lack divine guidance) in contrast with the Aristotelean model of a finite universe governed by a "prime mover." In an Aristotelean universe the fixed stars represented not only the end of the perceptual cosmos but also the end of all being. For the atomists, the multiple worlds whose existence they intuited lay beyond the fixed sphere of the heavens. Each world had its own surrounding spheres and stars. To be clear, this ancient understanding of infinite worlds is more closely related to contemporary notions of parallel universes then to exoplanet astronomy, insofar as their existence lies outside empirical study. Such other worlds could not be physically pointed to but served as the completion of thought experiments on the ramification of the atomists' theories. The conversation of other worlds was tied up in theological and metaphysical debates. Discussing the existence of these worlds, therefore, was a discussion simultaneously about Earthly and daily being. For Epicurus and his followers, their speculations on infinite worlds were inseparable from questions of theology and metaphysics. Stating the existence of other worlds was not primarily about answering the question of whether we are alone in the universe but was concerned with limiting the invocation of the divine as explanations for Earthly actions.

The demarcation of scientific questions from theological questions is of course a relatively recent occurrence. An example of one of the earlier attempts to present the latest understandings of the universe separate from theological conversation can be found in the writings of Bernard le Bovier de Fontenelle, who worked to convey the new knowledge of the cosmos to a popular audience during the French Enlightenment. In 1686 Fontenelle published *Entretiens sur la pluralité des monde*, which was translated into English in 1687 as *Conversations on the Plurality of Worlds*. It is written as a conversation between a teacher and a naïve woman, in a progressive attempt, Fontenelle explains in his introduction, to make these exciting ideas accessible to women as well as men. Fontenelle also addresses the religious reader, what he calls the audience "most difficult to satisfy" (1990, 5). To appease the person who would take offense to speculating that other planets are peopled, Fontenelle simply asserts that he is not referring to people as we know them but more broadly inhabitants who have not necessarily sprung from Adam and therefore do not violate biblical understandings. With this caveat stated, Fontenelle feels himself now freed to consider the nature of the universe from the exclusive perspective of the natural philosopher.

Conversations proceeds by explaining the heliocentric model of the universe, speculating on the Moon as an inhabited world, offering information on the other known planets, and finally suggesting that every "fixed star" is itself a Sun with orbiting worlds. On the broaching of this idea, the pupil exclaims, "Here's a universe so large that I am lost, I no longer know where I am, I'm nothing" (Fontenelle 1990, 63). This feeling of being lost is reiterated when the student considers that these worlds might each be inhabited: "We ourselves, to whom the same phrase [inhabitants] applies — admit that you'd scarcely know how to pick us out in the middle of so many worlds. As for me, I'm beginning to see the Earth so frighteningly small that I believe hereafter I'll never be impressed by another thing" (63–64). Such exclamations serve as a foil for Fontenelle, acting the wise teacher, to reframe such fear as excitement: "Now that they've given infinitely greater breadth and depth to this vault by dividing it into thousands and thousands of vortices, it seems to me that I breathe more freely, that I'm in a larger air, and certainly the universe has completely different magnificence. Nature has held back nothing to produce it; she's made a profusion of riches altogether worthy of her" (63).

The idea of many worlds, in this interchange, no longer concerns the role of divinity but has become a more personal question. How does it *feel* to think of other worlds: is it alienating or exciting? Fontenelle, through the voice of the tutor, clearly invites his Enlightenment audience to turn away from the fear of this enormity and embrace its possibility. Yet, the fear articulated by the student that Earth should seem inconsequential in the context of an expansive universe is not so easily countered. In chapter 4, I recalled Hannah Arendt's warning against humans extending not only their senses but their physical bodies into outer space. Nearly three centuries after Fontenelle's fictional student, Arendt similarly expresses the student's concern that the new way of being in the cosmos (or at the very least the solar system) proposed by proponents of the space age, this physical dislocation, would change what it means to be human as it would no longer be in reference to existence on Earth.

A few years after Arendt's essay, the Italian writer Italo Calvino published his short story collection *Cosmicomics*, containing a haunting story that echoes Arendt's concerns of terrestrial alienation. In "The Distance to the Moon," Calvino describes a time when the Moon was much closer to the Earth. The narrator of this story, accustomed to jumping back and forth between the two spheres, was startled to find himself stranded on the Moon as it quickly began drifting away from the Earth. During this unexpected exile, he realized how much he defined who he was with respect to his relationship to Earth: "I thought only of the Earth. It was the Earth that caused each of us to be that someone he was rather than someone else.... I was eager to return to the Earth, and I trembled at the fear of having lost it ... torn from its earthly soil, my love now knew only the heart-rending nostalgia for what it lacked: a where, a surrounding, a before, an after" (Calvino 1965, 14). When astronauts finally did leave Earth's surface and atmosphere, they became the first to understand what it would mean to return to Earth. From the grey, dusty Moon, astronauts talked about the beauty of the Earth, glowing blue in a sea of blackness. But even as a small portion of humans became spacefaring, the fear that Earth would diminish into an inconsequential speck rang false. Rather, one may argue that the view of Earth from space did just the opposite. Earth, seen as a planet, became an important rhetorical image for the dawning of ecological consciousness.

The most prominent legacies of the space age are not prolonged human presence in space and exploration of nearby planets but a new way to observe and study our own planet.

Astronomers and planetary scientists are often in the position of justifying their research. Why should we care about other worlds when there are pressing problems on our own world? Some Mars scientists, as I showed in chapter 1, employ the rhetoric of the "frontier" to argue that freedom of expression and the generation of new ideas depends on maintaining an open and other landscape of exploration. Connecting research to respected thinkers, be they Turner or Epicurus, acts as a justification of importance. Today's work continues to address questions important to previous generations. In other words, these questions have already been deemed important and since they remain unanswered should "obviously" continue to be studied.

But even with these appeals to the past, there is a different cosmology underlying the current study of other planets. Scientists do not see their work as explicitly bolstering a theological stance, as did Epicureans, nor do they need to offer comfort to a population still adjusting to a post-Copernican worldview, as Fontenelle felt compelled, or offer provocation to a population anticipating human space flight, as Arendt and Calvino did. In fact, whereas the student in Fontenelle's *Conversations* is disoriented by learning of the possibility of multiple worlds, today's exoplanet astronomers are suggesting that to *not know* of these worlds would be alienating. In March 2009, just before the Kepler satellite's launch, a *New York Times* article asserted the significance of finding an Earth-like planet this way: "Someday it might be said that this was the beginning of the end of cosmic loneliness" (Overbye 2009b). That knowing of the existence of other planets will make humans feel less alone is what distinguishes the contemporary conversation on the plurality of worlds from previous iterations. It is also why placemaking figures so prominently. We will only be less alone if we can connect with, imagine *being* on, and thus being in place on, these other worlds.

Scholars have cited increased loneliness and solitude as a by-product of the changing role technologies are playing in daily life (notably Putnam 2000 and Turkle 2012; see also Hampton et al. 2015 for a refutation). Yet exoplanet astronomers are arguing that it is through the grand technological feat of detecting a world like our own that humans will finally feel less cosmically alone. The question of "our place in the universe," astronomers claim, will finally be answered. As I discussed in chapter 4, this answer will come in the form of a detection of a habitable, Earth-like planet. This is an astronomical object made into a very specific kind of place—one whose surface and atmosphere are imagined in great detail, even if the life that might lurk on that surface remains indeterminate. This is the planet to which astronomers hope to one day be able to point; to connect our terrestrial way of being with a cosmic way of being. This, they claim, will neither aggrandize nor diminish Earth but will allow us to finally know Earth, for we will come to know how Earth relates to other planetary places.

Regardless of the time or the reigning cosmology, then, speculating on the plurality of worlds provokes thinking not only about the universe but about Earth itself. The entanglement between worlds, what I described as a multiple exposure in chapter 1, is present in each subsequent chapter and indeed in the references I have made in this conclusion. Ideas of what it means to be on Earth shape studies of other planets, and studying the habitability of other worlds refines how we define life on Earth. Place draws together this intergalactic network, serving as a metric of meaning-making and analogy.

To speak of a "planet" is never to speak of an isolated body, as today's planetary scientists are constantly configuring connections and comparisons. As I discussed in the introduction, some social scientists are turning to an idea of the planet or planetarity as an alternative to writing about "globalization." Planet, in these conversations, refers to an Earth divorced from its broader cosmological context. I have already suggested that attending to the ways planetary scientists invoke place at a planetary scale might invite a sense of embodiment even when thinking about phenomena of a global scale, thus specifying social scientific invocations of "the planetary." I would also like to bring a comparative planetarity into this discourse. What questions might we raise if we take seriously the astronomical claim that to know Earth and to know ourselves requires that we know other worlds? The planetary imagination is fluid and changing, just as our own sense of Earth as a planet and a place must be. We can learn from planetary scientists that these changes are part of a broader, universe-spanning cosmology. Placing outer space draws a new cosmos, but also points to how other worlds matter for being on this world.

Introduction

1. "Planetary science" traditionally refers to the study of planets within our solar system. This discipline formed before exoplanets were discovered, and it is increasingly common for those trained in studying solar system planets to apply their methods to exoplanets.

2. Icarus was given a pair of wings made of feather and wax by his father, Daedalus. Though Daedalus warned Icarus not to fly too high, he was swept up in the joy of soaring through the sky and eventually flew too high and too close to the Sun. The wax that bound the wings together melted, and Icarus met a watery death in the ocean below.

3. For more on the history of planetary science in America, see Doel (1996), Chamberlain and Cruikshank (1999), and Launius (2013).

4. Statistics generated using the Web of Science database, searching across *Astrophysical Journal*, *Astrophysical Journal Letters*, and *Astronomical Journal*. The number of articles reported is based on a search for "exoplanet" and "extrasolar planet" in topic.

5. Though, as I put forth in the previous section, "planet" is both cultural and natural.

6. At the same time that Latour invites us to think of Earth as an actor, he also suggests that we need not think beyond the Earth. See Olson and Messeri (2015) for a critique of this position.

7. For more on "the planetary," see Morgan (1984), Pratt (1992), Friedman (2010), Masco (2010), Jazeel (2011).

8. Two localities most often associated with scientific work are the laboratory (Latour and Woolgar 1986; Traweek 1988; Knorr-Cetina 1999; Silbey and Ewick 2003) and the field (Haraway 1990; Kuklick and Kohler 1996; Hayden 2003; Lowe 2006; Helmreich 2009). Each site holds a different epistemological promise, with the lab taking on an aura of placelessness, and thus universalism, and the field being valuable precisely because of where it is located (see Kohler 2002). While laboratory studies can very well be thought of as a subdiscipline within science and technology studies, studies of the field sciences have been less focused on spatial matters. Often it has been sufficient to remark on how field scientists import elements of laboratory work in order to legitimize their emplaced practice (see, for example, work on the "personal equation" by Schaffer 1988; Canales 2002b), reinforcing rather than questioning the relationship between "placelessness" and "objectivity." More recent studies in science and technology studies and anthropology of science have challenged the lab/field dichotomy (see, for example, Hayden 2003). Further, though scientific practice is local and situated, conceiving of the place of science in terms of the field or the lab limits our understanding of how knowledge travels. These sites are one node in a wider, dynamic network (Latour 1987; Secord 2004; Kaiser 2005b). More recent studies of the spatiality of science suggest that it is the relationship between the laboratory and the field that is crucial to understanding the production of scientific knowledge (Henke 2000; Kohler 2002; Livingstone 2003; Gieryn 2006).

9. The social studies of outer space includes historians and sociologists who study astronomy (Edge and Mulkay 1976; Pinch 1986; Schaffer 1988; Lankford 1997; Canales 2002a; Stanley 2007; Munns 2012), the political and social history of human space flight (Logsdon 1970; McDougall 1985; Siddiqi 2000; Ackmann 2003; Mindell 2008; Bimm 2014; Rand 2014), the ethnography of space institutions and projects (Mack 1990; Zabusky 1995; Vaughan 1996; Redfield 2000; Mirmalek 2008; Olson 2010; Valentine 2012; Vertesi 2015), and the cultural meaning of extraterrestrial narratives and phenomena (Young 1987; Lepselter 1997; Dean 1998; Battaglia 2006; Lempert 2014).

10. I note here the naming conventions used throughout this book. I refer to many of the scientists by their real names with their permission. These are all senior, well-established researchers. Other senior researchers I anonymize either because they have asked or their identity is inconsequential to my argument. I always provide junior scholars, even when permission was given to use real names, with pseudonyms. The convention of this text is that informants named in full or referred to by their last name are real names, while informants for whom I provide only a first name are pseudonymous.

Chapter One. Narrating Mars in Utah's Desert

1. Every crew that visits MDRS, from the first crew in 2002 through to the writing of this book, is responsible for submitting multiple daily reports. They are posted on the public website of the Mars Society (http://MDRS.marssociety.org/crew-reports).

These reports range across timelines of the day's activity, scientific and engineering reports, and more reflective journal entries. Not all of these reports are archived; most of the reports from 2007–2011 are missing.

2. For more on the Mars Underground, see Chaikin (2008, 123–53).

3. Stoker served on the Mars Society steering committee for the first ten years of its existence.

4. A more volatile account of the split between NASA and the Mars Society is written in a memoir by Zubrin, citing conflict of personalities and discrepancy over financing between him and the NASA principal investigator, Pascal Lee (R. Zubrin 2004, 260–63).

5. Stoker is collaborating on this project with Bernard Foing from the ESA. Each crew going to MDRS for this project was intended to reflect this international collaboration.

6. Stoker requested I compile the data I collected into a poster, which was presented at the National Lunar Science Forum in July 2010 (Messeri, Stoker, and Foing 2010).

7. This has been accomplished in a variety of ways by positioning landscapes as modes of production (Lefebvre 1974; Harvey 1989; Zukin 1993), as gendered (Rose 1993), and as sites of contestation (Duncan and Duncan 1988; Duncan 2005; Mitchell 1996). For a general overview of writings on landscape in geography, see Mitchell (2005). Denis Cosgrove (1985) was particularly influential in showing that even the idea of landscape (in the colloquial sense of landscape painting) is a product of a particular time, when geometric perspective was first learned and taught. Thus, the idea of landscape embodies the power dynamics of that moment. Landscape, he concluded, is a way of seeing, and the attending visual ideology was bourgeois, individualist, and exercised power over space. For Cosgrove, landscape is both a social product and a detached way of structuring the world.

8. Entrikin (1991) has linked narrative and place-time by extending Bakhtin's concept of the "chronotope" (1981) into human geography.

9. For more on the mutual construction of landscape and identity, see Fred Myers's (1993; 2000) work on aboriginal Australians.

10. Interview of Don Wilhelms by Ronald Doel, June 22, 1987, Niels Bohr Library and Archives, American Institute of Physics, College Park, Maryland, available at the website of the Institute, https://www.aip.org/history-programs/niels-bohr-library /oral-histories/5064.

11. Rankama might perhaps be overly sensitive to the use of geological terms in other fields, as he ends his article with an admonishment of the petroleum engineer who has inappropriately "snatched" terms from petrologists and geochemists. In the early 1960s, geology appeared to be struggling with its boundaries, as was echoed in a presidential address to the Geological Society of America delivered by M. King Hubbert (1963). Hubbert expressed discomfort with the disproportionate amount of geologists entering gainful employment in the petroleum industry, an industry he

predicted was soon to decline and no longer be a source of employment for geologists.

12. Interview of Michael Carr by Ronald Doel, June 22, 1987, Niels Bohr Library and Archives, American Institute of Physics, College Park, Maryland, available at the website of the Institute, https://www.aip.org/history-programs/niels-bohr-library/oral-histories/5088.

13. Interview of Eugene Shoemaker by Ronald Doel, September 8, 1988, Niels Bohr Library and Archives, American Institute of Physics, College Park, Maryland, available at the website of the Institute, https://www.aip.org/history-programs/niels-bohr-library/oral-histories/5082–4.

14. Frederick Turner (1893) famously argued that the frontier shaped American notions of individuality and democracy. There is a significant literature produced by U.S. historians revisiting, revising, and critiquing the Turner thesis. I will simply refer to Patricia Limerick's work (1988), both because she offers a nuanced history of the American West and because she has elsewhere (Limerick 1992) written against the use of the frontier metaphor in the space program.

15. DeGroot shows how the frontier spirit animated NASA as early as the Mercury program. The first astronauts spoke of themselves as pioneers ready to explore the frontier of space. DeGroot nicely suggests that, in the wake of the atomic bomb and the shattering of the romance of war, the press and public welcomed the resurrection of the fantastical frontier explorer (2006, 109). See also Parker (2009, 89–91) for a discussion on how capitalism intersects with the frontier narrative in the space program.

16. Interview of Eugene Shoemaker by Ronald Doel, September 8, 1988.

17. "About" page, website of the Association of Mars Explorers, accessed January 17, 2016, http://mars-explorers.org/about/.

18. The author of this entry is Maggie Zubrin, Bob Zubrin's wife.

19. Elon Musk, contradictory advocate of clean transportation technologies (Tesla Motors) and mass immigration to Mars and commercial space flight (SpaceX), came across the Mars Society in 2001 when Zubrin was raising funds to build MDRS. Musk donated $100,000, and the small observatory was named in his honor; see Ashlee Vance "Elon Musk's Space Dream Almost Killed Tesla," May 14, 2015, Bloomberg Business, http://www.bloomberg.com/graphics/2015-elon-musk-spacex/.

20. See also Jameson (2005) for a discussion on utopia and Robinson's trilogy.

21. I would be remiss if I did not mention Jameson's formulation of utopia and science fiction and how it operates in some ways opposite to what Markley suggests. In writing about these literary genres, Jameson (1982) argues that in studying these works one comes to the ironic conclusion that they do not represent the future but instead draw attention to our inability to conceive of utopia. They (unsurprisingly) illuminate the present moment as opposed to a past or future.

1. I spent six months with the Mapmakers, February through August 2010.

2. See, for example, Pratt (1992) on the role of maps in developing a European "planetary consciousness."

3. I want to be clear that the mapping projects I observed were undertaken with genuine commitment and enthusiasm for offering people a "cool" experience of outer space. The Mapmakers do not see their work as necessarily political even if my analysis draws attention to the ways it is inherently so.

4. This history of Ames comes from the resident historian, Glenn Bugos (2000). Glenn was a great source of help while I was at Ames, helping me navigate the archive and suggesting people to speak with.

5. NASA was established in 1958 by the National Aeronautics and Space Act. It outlines provisions under which NASA can arrange contracts with other entities. For partnerships not explicitly discussed in the Space Act, Space Act Agreements are drawn up that stipulate legally enforceable commitments.

6. Website at pds.nasa.gov. Last updated January 2016.

7. The problem with this invocation of democracy, with regard to unequal access to Web 2.0 cartographies, is discussed in Crampton (2009), Farman (2010), and Haklay (2013).

8. MOC was mounted on the Mars Global Surveyor, which launched in 1996. This was the first successful American mission to orbit Mars since Viking 2 in 1975. Launched one month after Surveyor, Mars Pathfinder was the first robot to land on Mars that could traverse the surface. Despite several failed missions, Mars satellite and robotic exploration remained active throughout the 1990s and 2000s. The two Mars Exploration Rovers, launched in 2003, operated on the Martian surface for more than five years.

9. Kelty deftly works through the differences (and similarities) between free software and open source and the evolving relationship between these two terms and their attending communities. He explains that they are two movements with the same material base but different ideologies. See specifically his chapter 3. I use "open source" in my discussion because that is the language used by the Mapmakers.

10. Anthropologist Gabriella Coleman (2004) similarly discusses the buried politics of open source and free software movements. She characterizes this community as having a "political agnosticism," meaning that even as programmers deny any political allegiance, their coding initiatives belie an ethic devoted to free speech and transparency. Elsewhere, Coleman and Alex Golub (2008) articulate the hacker ethic as a particular kind of liberalism. The ways the Mapmakers invoke democracy resonates with their use of liberalism.

11. One Mapmaker corrected my description of "unquestioned." In the past he, and he assumed others in the group, made the conscious choice that openness was a

good thing to pursue in his professional and personal life. Just because the community no longer questions the goodness of open source, does not mean the individual did not question this choice before joining the initiative.

12. The Mapmakers also do significant work with lunar data. Their first release of an interactive lunar map for Google was the website www.google.com/moon, which went live in 2008. The Moon in Google Mars was released on June 20, 2009, to commemorate the fortieth anniversary of *Apollo* 11 landing on the Moon. For the purpose of this chapter, though, I focus on the work they do with Mars images.

13. See Harley's influential "Deconstructing the Map" (1989) for more on how all maps, even scientific maps, can be "read" for the metaphors and symbols they contain.

14. HiRISE has a program called HiWish, which allows non–team members to suggest surface locations to photograph. This has led to NASA officials referring to HiRISE as "the people's camera" and HiWish as "participatory exploration." The democratic ethos spreads beyond the maps, a point to which I will return in the conclusion.

15. The footprint of the HiRISE images are included in Mars in Google Earth, but this only provides an outline of where the image fits and a link to the geographically decontextualized HiRISE image as it is available on the Internet.

16. The scheme that TOAST uses to model a sphere as a polyhedron is the hierarchical triangular mesh. This is based on an earlier solution to the problem called the octahedral quaternary triangular mesh. The geographer Geoffrey Dutton, who invented this model, acknowledges that he was inspired by Fuller's Dymaxion projection and the geodesic dome (Dutton 1996).

17. "NASA and Microsoft provide Mars 3-D Close Encounter" by Michael Mewhinney and Rachel Hoover, last accessed July 2010, http://www.nasa.gov/home/hqnews/2010/jul/10–163_Microsoft_Mars.html.

18. A DVD of this film was given to me by Glenn Bugos at the NASA Ames History Office.

19. Lederberg, "Stanford's Trip to Mars," address, Medical Alumni Reunion Day, May 22, 1976, PP04.02, Elliott C. Levinthal Viking Lander Imaging Science Team Papers, 1970–1980 (hereafter NASA ARC, PP04.02, Levinthal Collection), Box 11, Folder 59, NASA Ames History Office, NASA Ames Research Center, Moffett Field, California. See also Wolfe (2002) for Lederberg's role in constructing exobiology as a peaceful and scientific venture, and Dick and Strick's (2004) account of NASA's role in the rise of exobiology and the subsequent transformation into astrobiology.

20. "Prof Makes 3-D Photos of Mars," n.d., NASA ARC, PP04.02, Levinthal Collection, Box 9, Folder 50.

21. Elliott Levinthal to Geoffrey Briggs, March 2, 1979, NASA ARC, PP04.02, Levinthal Collection, Box 9, Folder 46.

22. Byron Morgan to Elliott Levinthal, October 1, 1980, NASA ARC, PP04.02, Levinthal Collection, Box 9, Folder 81.

23. The Viking Fund was proposed by Eric Burgess, who also had the idea that the Pioneer probes should carry messages about their origin. After Burgess pitched this idea to Carl Sagan, the Pioneer plaque was designed (see Vakoch [1998] for an analysis of the symbolism of the plaque).

24. Bill Copeland to Elliott Levinthal, n.d., NASA ARC, PP04.02, Levinthal Collection, Box 9, folder 42.

25. Following Viking, no space program launched a successful mission to Mars until NASA's Mars Global Surveyor and Mars Pathfinder missions in 1996. In 1989 Russia lost two orbiters and landers en route to or shortly after orbital entry around Mars. In 1992 NASA had a similar failure with Mars Observer.

26. Film scholar Thomas Elsaesser (2010) historicizes the current iteration of 3-D cinema and why digital 3-D does not rectify the earlier problems faced by 3-D movies. See also Elsaesser (2013).

27. The model is created by first finding corresponding coordinates between two images of roughly the same terrain. With a relationship established between the two images and knowledge of the camera position, a point cloud can be generated. A point cloud is a grid of locations over a surface that provides x, y, and z coordinates for each point. The next step is to convert the point cloud into a polygonal "mesh." A mesh draws lines between the points in the point cloud, transforming a discrete grid into a series of triangles. The surface is now connected and can be visually rendered by most 3-D browsers. From the point cloud, Stereo Pipeline also can output a digital elevation model, which is the quantitative result of the 3-D processing and useful for scientists who want to do more than just look at their 3-D creations.

28. In addition to wanting the partnership he forged to succeed, Kemp was invested in the Microsoft project because it was a high-profile use of NASA's cloud computing platform, Nebula, which was one of Kemp's main initiatives.

29. This scheme of eras was developed at the USGS in the mid-1980s. Each era is named for the region exemplifying the presumed geology of that time. On Martian nomenclature, see Greeley and Batson (1990, 103–7).

30. Personal correspondence with Jim Garvin, March 18, 2010.

Chapter Three. Visualizing Alien Worlds

1. The animation was created by graphic artist Dana Berry.

2. A handful of exoplanets have been directly imaged. This is done by blocking out the light of the star such that the dim exoplanet has opportunity to shine through. In these cases, the photographed exoplanets appear as points of light. Their surface composition cannot be discerned.

3. Exoplanet astronomy is not the only scientific field that visualizes unseen phenomena. See Dumit (2004), Joyce (2005), Collins (2004), and Galison (1997) for studies that explore various facets of invisible phenomena in the areas of brain imaging, gravity wave detectors, and bubble chambers.

4. See Pina-Cabral (2014) for an attempt to pin down what anthropologists mean when writing about worlds. For other anthropological writings that explicitly engage with questions of worlds and worldings, see Helmreich (2000), Zhan (2009), Descola (2010), and Tsing (2010).

5. Statistics generated using Web of Science topic search for "exoplanet" and "extrasolar planet." The increase corresponds to the number of exoplanets discovered. Dedicated exoplanet telescopes and projects that were established starting around 2006 account for this dramatic increase.

6. For science and technology studies thinking on pedagogy, see Kaiser (2005a). See also Grasseni (2009) for ethnographic accounts of various tacit techniques by which modes of vision are adopted by communities.

7. The names of exoplanets are alphanumeric; a lowercase letter affixed to the star name. The star name is derived from its catalog number (for example HD 209458) if it does not have a formal name (as most stars do not). In other cases, the star is named for the survey that detected it. In this instance, CoRoT-N is the name of the star, and CoRoT-Nb is the first planet discovered around that star. Subsequent planetary discoveries around the same star would be subtitled c, d, etc.; a is never used to denote a planet because that letter implicitly stands for the star itself.

8. The extent to which stars are variable on this time scale also had implications for the Kepler satellite (discussed in chapter 4). Seager sought to confirm that stars were less variable on shorter time scales.

9. The distinction between visual and photographic photometry is an actor's distinction. Whereas seeing with the eye through the telescope was considered a direct observation, when these techniques were first being pioneered, photographic evidence was not considered to be direct observation (as evidenced by Scheiner [1894, 2]). Now, the distinction between direct and indirect is not related to human versus mechanical detector. Direct means to measure the planet; indirect means measuring the star.

10. This process of calculating the magnitude from a photographic plate was called "reducing," the same word used today to describe the elimination of environmental and instrumental systematics from the data.

11. NASA defunded NStED and now supports the Exoplanet Archive to house public exoplanet data.

12. For discussions of tacit knowledge in scientific practice see Polanyi (1958) and Collins (1974). While these scholars focus on the materiality of practice and how technicians teach people how to use specific instruments, other scholars have considered how tacit knowledge remains important in nonmaterial, theoretical work (Kennefick 2000; Warwick 2003; Kaiser 2005b).

13. The drama surrounding the Gliese 581 system continued for several years after this meeting. Vogt, at a conference in 2011, provided new data and once again claimed that 581g existed and indeed existed within the habitable zone. However, an article published in *Science* in 2014 once again questioned this claim (Robertson et al.

2014). This article, which as of this writing reflects the exoplanet community consensus, argues that the planetary signals of both Gliese 581g and another believed planet in the system, 581d, were artifacts of stellar activity.

14. Though my norm has been to use pseudonyms for Seager's students, I use Madhu's real name. He is now a university lecturer in astrophysics at the University of Cambridge, and the article he authored that I draw from in this section has had great impact, cited by more than two hundred scholars.

15. Spectra for exoplanets are generated when the radiation from a much hotter host star passes through the much cooler atmosphere of an exoplanet during an observed transit. The molecules and atoms of the exoplanetary atmosphere absorb the stellar radiation as it passes through. A planetary spectrum, then, is made by subtracting the spectrum of the star from the spectrum of the planet and star.

16. Seager, Madhu's dissertation advisor, suggested the "million model approach," and Madhu found an elegant solution. The credit I give to Madhu in this section also deservedly belongs to Seager as his mentor.

17. A further note on what is represented in the TP profile. It is a graph of temperature (on the horizontal axis) versus pressure (on the vertical, in descending value). As pressure increases the further you move from the surface, a TP profile represents how the temperature changes as you move from the surface into the atmosphere. For planets in the solar system, scientists obtain the TP profile based on a detailed spectrum. For exoplanets, no detailed spectrum is available. Analysts have attempted to solve this problem by deriving a range of TP profiles and molecular and atomic abundances based on a given spectrum by solving three equations simultaneously. This method is too computationally intense to run multiple times. Madhu proposed a parametric TP profile that would satisfy hydrostatic equilibrium and global energy balance equations. He wrote an exponential equation that proved a good fit for the solar system planets (Seager 2010). With this self-contained equation, containing six free parameters, Madhu was able to run many thousands of possibilities on the same data.

18. The nomenclature GJ refers to a catalog containing known stars within twenty-five parsecs of Earth. Wilhelm Gliese and Hartmut Jahreise prepared and updated the catalog at the University of Heidelberg, hence the GJ honorific.

19. As used here, "interior" refers to the core, mantle, and gas envelope. The gas envelope is the gaseous layer surrounding a rocky concentration of mass.

20. Though Jessica did explore the possibility of pursuing a project on stellar evolution, in a quest for more definite answers, she returned to interior modeling for her dissertation work. Her research continued to be motivated by the work she did with GJ 1214b, proposing circumstances under which super-Earths might have liquid water on their surface (personal email to author, July 11, 2011).

21. I did not participate in the research or the writing of this article, but my opinion was sometimes asked for as Jessica and Seager tried to figure out how best to express a particularly sticky point.

22. Medical anthropologist Barry Saunders (2009) illustrates how learning to see and speak are coupled in the world of diagnostic medicine.

23. Astrometry, like the Doppler method, detects planets by observing the motion of the parent star. Whereas the Doppler method measures the speed of the star, astrometry measures side-to-side motion of the star's position in the sky. Though astronomers have been using astrometry since the nineteenth century to study binary star systems, it has thus far been unsuccessful in detecting exoplanets. In the mid- to late twentieth century, several planetary detections were made using astrometry, but all were later declared erroneous (Boss 2009).

24. It seems the group skepticism was warranted. Less than a year after this discovery, an article came out by a different group claiming that no variability was detected in the host star and the prior planetary claim was unfounded. The Exoplanet Encyclopedia (exoplanet.eu) catalogs this "planet" under "unconfirmed, controversial, or retracted planets."

Chapter Four. Inhabiting Other Earths

1. In all three versions of this story that I have told, the two in the introduction and the one here, the scientist is always a woman. In a personal correspondence with Lunine, he said that he had several inspirations in mind for the mother, including Sara Seager, whose work I discussed in detail in chapter 3, and Debra Fischer, to whom my attention turns in this chapter. That a mother, as opposed to a father, is featured serves to reinforce the implications of finding an Earth-like planet: Mother Earth might have multiple, life-nurturing kin. I do not take it as a reflection that exoplanet astronomy is a field that, unlike other physics subdisciplines, has a more balanced gender ratio. Seager (2013) wrote an opinion piece for the *Huffington Post* about the invisible structures that continue to favor the advancement of men within astronomy.

2. Shortly after Lunine's testimony, NASA did in fact pull the funding from this mission, which was called the Terrestrial Planet Finder. See Messeri and Vertesi (2015) for more on the Terrestrial Planet Finder as a perpetually deferred technology.

3. Kepler experienced a hardware malfunction in 2013 and was no longer stable enough to collect starlight from this sample. The telescope was repurposed for other science experiments that required a lesser pointing accuracy.

4. Like the authors of many works preceding this one, I am discussing observation in the context of astronomy. Lorraine Daston and Elizabeth Lunbeck (2011) have edited a volume that considers the history of observation across disciplines. Their metaphorical notion that "observation discovers the world anew" (1) is quite appropriately borne out in the search for habitable exoplanets.

5. For a history on the role of observatories in modern astronomy, see McCray (2004). For the social history of observatories, with an emphasis on gender roles, see Pang (1996), as well as Nisbett (2007), who addresses themes of gender and national

interest alongside the economic history of establishing observatories. Several volumes in the history of astronomy focus on astronomy beyond the observatory in the form of expeditions, notably nineteenth-century solar eclipses (Canales 2002b; Pang 2002).

6. The determination of too much or too little energy is presumably based on terrestrial conditions. Huang provides a circular justification. From his definition of habitable zone, he concludes that main-sequence F, G, and K stars are the most likely to host planets capable of life. "It is interesting to note," he concludes, "that our sun, which is a main-sequence G2 star, does support life abundantly on at least one of its planets, fully in agreement with the present conclusion." Also in this article, Huang speculates that if planets existed around Alpha Centauri (Fischer's star of interest), it is unlikely that they could orbit completely within the habitable zone due to the dynamics of the multistar system.

7. In October 2012 the European team announced the detection of a planet (albeit uninhabitable) around Alpha Centauri B. This exoplanet was greeted with more fanfare than most: not only was it a planet around our closest star; it was also reportedly Earth-sized. Following this announcement, the Planetary Society, a nonprofit research group that has provided some support for Fischer's Project Longshot, posted Fischer's response to the discovery: "The indication that our nearest neighbor has rocky planets is incredible.... This leaves open the possibility of a terrestrial planet in the habitable zone—in fact, I think this strengthens the speculative possibility of a habitable world in the alpha Cen system" (Betts 2012). In the years following this discovery, Fischer and her team tried (and failed) to confirm this discovery. In another project update posted by the Planetary Society, she questioned the existence of this planet and reiterated that even more sophisticated equipment will be needed to study the Alpha Centauri system (Betts 2014). In the fall of 2015, another team confirmed Fischer's doubt and proved that the "planet" was an artifact in the data (Rajpaul, Aigrain, and Roberts 2016). This paper convinced the community that no planet is yet known to exist in the Alpha Centauri system.

8. This history of CTIO comes from two accounts written by Victor Blanco, the observatory's director from 1967 to 1981 (Blanco 1993; 2001).

9. Though I mention Marcy only briefly, he is one of the founding figures of American exoplanet astronomy. In the fall of 2015, the University of California, Berkley found Marcy in violation of the school's sexual harassment policies following a Title IX investigation. The outcome of this investigation was widely covered in the news. On social media the hashtag "#astroSH" brought attention to Marcy's actions and prompted a larger conversation about the challenges faced by women in astronomy. Ultimately, the community reaction was more condemning than the findings of the Title IX investigation. Marcy announced his intention to retire as a consequence of this reaction.

10. The mule trail only existed because when Jürgen Stock began surveying the area in 1960 he hired local workers to build a path to Tololo's summit. Before Stock

began frequenting Tololo, it is unclear whether there was a formal path to the top. This and other details from Stock's expedition are recorded in the Stock Reports, letters written from Stock to his supervisor, Donald Shane, at the Lick Observatory, now held in the Observatory's library. I thank Ana Veliz, the CTIO librarian, for scanning and emailing me these reports.

11. Stock Report No. 1, 1960, pg. 1.

12. Stock Report No. 10, October 30, 1960, pg. 2.

13. Chilean astronomers were involved in CTIO planning from the beginning. A Santiago astronomer, Carlos Torres, was Stock's second in command. However, a few comments in Stock's letters make it clear that Stock never viewed Torres as a colleague.

14. Victor Blanco (1993), the second director of CTIO, does mention that fifteen families lived on the land AURA purchased to build CTIO. They were goat herders, so in order to keep the observatory's water source clean, the Ramos family, in particular, had to be displaced downstream. Blanco remarked that moving this family "created a difficult human-relations problem" that was resolved by making the family the land managers of the area surrounding CTIO (excluding the summit). At some point, the Ramos family turned their house into a bar-discotheque, which resulted in increased nighttime traffic on the Tololo road. This situation "was promptly terminated" as it severely interfered with observing conditions. The other goat herders whose animals did not interfere with the water supply were charged rent but allowed to remain on the land.

15. The purpose of this particular run was replacing the old iodine cell with a new one, so there was more interaction with the instrument than usual. Fischer hand-delivered the cell, wrapped carefully in foam, and spent some time during the nights swapping the new cell in and out for calibration purposes.

16. Stefan Helmreich (2007) offers an alternative to thinking about fieldwork. Rather than "immersive," it might be fruitful to think of ethnographic work as "transductive." He develops this rubric for his own approach to an anthropology of sound but suggests that transduction attends to how presence coalesces. Whereas the subject and field are already merged in an immersive approach, the transductive approach draws attention to how this immersion came to be. For example, what are the material and sensorial means by which the astronomers (and myself as an anthropologist) enter a state of "being there" at the observatory?

17. The frustration I note in the Kepler project surrounds the individual's access to data, the positioning of the "I" among the collective "we." Peter Galison (2003) has tracked the changing position of the collective we, and attending changes to claims to knowledge, as instruments and projects have grown in scale. Sharon Traweek (1988) has shown how assumptions as to how groups produce scientific facts are embedded within the design of particle accelerators. The "we" of science is apparent in several other studies of large-scale collaborations, including Harry Collins's (2004) work on gravitational waves and Galison's on the bubble chamber (1997).

18. Many scholars have written commentary and critique on the images of Earth taken during the Apollo missions. For an overview of the iconography, see Cosgrove (1994). Jasanoff (2004) discusses how the environmental movement harnessed these images. Garb (1985) suggests alternate readings for Earth images.

19. See also Rose (1993) and hooks (1990).

Ackmann, Martha. 2003. *The Mercury 13: The Untold Story of Thirteen American Women and the Dream of Space Flight*. New York: Random House.

Aigrain, S., F. Pont, F. Fressin, A. Alapini, R. Alonso, M. Auvergne, M. Barbieri, et al. 2009. "Noise Properties of the CoRoT Data: A Planet-Finding Perspective." *Astronomy and Astrophysics* 506 (1): 425–29.

Arendt, Hannah. 1963. "The Conquest of Space and the Stature of Man." In *Between Past and Future: Eight Exercises in Political Thought*, 265–82. New York: Penguin.

Auvergne, M., P. Bodin, L. Boisnard, J. T. Buey, S. Chaintreuil, G. Epstein, M. Jouret, et al. 2009. "The CoRoT Satellite in Flight: Description and Performance." *Astronomy and Astrophysics* 506 (1): 411–24.

Bachelard, Gaston. 1958. *The Poetics of Space*. Translated by Maria Jolas. Boston: Beacon Press.

Bakhtin, Mikhail Mikhaïlovich. 1981. *The Dialogic Imagination: Four Essays*. Translated by Caryl Emerson and Michael Holquist. Austin: University of Texas Press.

Barge, P., A. Baglin, M. Auvergne, H. Rauer, A. Léger, J. Schneider, F. Pont, et al. 2008. "Transiting Exoplanets from the CoRoT Space Mission." *Astronomy and Astrophysics* 482 (3): 17–20.

Basso, Keith. 1996a. *Wisdom Sits in Places: Landscape and Language among the Western Apache*. Albuquerque: University of New Mexico Press.

Basso, Keith. 1996b. "Wisdom Sits in Places: Notes on a Western Apache Landscape." In *Senses of Place*, edited by Steven Feld and Keith Basso, 53–90. Santa Fe, NM: School of American Research Press.

Batalha, Natalie. 2013. "Exoplanets and Love: Science That Connects Us to One

Another." Interview with Natalie Batalha by Christa Tippett. *On Being.* NPR, August 29.

Battaglia, Debbora, ed. 2006. *E.T. Culture: Anthropology in Outerspaces.* Durham, NC: Duke University Press.

Baudrillard, Jean. 1983. *Simulations.* Translated by Paul Foss, Paul Patton, and Philip Beitchman. New York: Semiotext(e).

Beattie, Donald A. 2001. *Taking Science to the Moon: Lunar Experiments and the Apollo Program.* Baltimore: Johns Hopkins University Press.

Betts, Bruce. 2012. "First planet discovered in Alpha Centauri system." The Planetary Society Blog, October 18, http://www.planetary.org/blogs/bruce-betts/20121017 -Alpha-Centauri-first-planet-discovery.html.

Betts, Bruce. 2014. "Update on the search for planets in the Alpha Centauri system." The Planetary Society Blog, April 4, http://www.planetary.org/blogs/bruce-betts /20140401-update-on-the-search-for-planets.html.

Bimm, Jordan. 2014. "Rethinking the Overview Effect." *Quest* 21 (1): 39–47.

Blanco, Victor. 1993. "Brief History of the Cerro Tololo Inter-American Observatory." Cerro Tololo Inter-American Observatory, accessed January 24, 2016, http://www .ctio.noao.edu/noao/content/CTIO-History.

Blanco, Victor. 2001. "Telescopes, Red Stars, and Chilean Skies." *Annual Review of Astronomy and Astrophysics* 39:1–18.

Borucki, W. J. 2010. "Brief History of the Kepler Mission." Kepler: A Search for Habitable Planets, May 22, http://kepler.nasa.gov/Mission/QuickGuide/history/.

Borucki, W. J., E. W. Dunham, D. G. Koch, W. D. Cochran, J. D. Rose, D. K. Cullers, A. Granados, et al. 1996. "FRESIP: A Mission to Determine the Character and Frequency of Extra-solar Planets around Solar-Like Stars." *Astrophysics and Space Sciences* 241 (1): 111–34.

Borucki, W. J., D. G. Koch, E. W. Dunham, and J. M. Jenkins. 1997. "The Kepler Mission: A Mission to Determine the Frequency of Inner Planets near the Habitable Zone of a Wide Range of Stars." In *Planets Beyond the Solar System and the Next Generation Space Missions*, edited by David Solderblom, ASP Conference Series, vol. 119: 153–73. San Francisco: Astronomical Society of the Pacific.

Borucki, W. J., and A. L. Summers. 1984. "The Photometric Method of Detecting Other Planetary Systems." *Icarus* 58 (1): 121–34.

Borucki, W. J., and A. T. Young. 1984. *Proceedings of the Workshop on Improvements to Photometry.* Vol. CP-2350. San Diego, CA: NASA.

Boss, Alan. 2009. *The Crowded Universe: The Search for Living Planets.* New York: Basic Books.

Boyd, Richard. 1993. "Metaphor and Theory Change: What Is 'Metaphor' a Metaphor For?" In *Metaphor and Thought*, 2nd ed., edited by Andrew Ortony, 481–532. Cambridge: Cambridge University Press.

Bugos, Glenn. 2000. *Atmosphere of Freedom: Seventy Years at the NASA Ames Research Center.* NASA SP-2010-4314. Washington, DC: NASA History Office.

Burbank, Sam. 2002. "Personal Journal." Mars Desert Research Station Crew Reports.

April 24, http://web.archive.org/web/20051119154449/http://www.marssociety
.org/MDRS/fs01/0424/samjourn.asp.

Bush, Vannevar. 1945. *Science, the Endless Frontier: A Report to the President.* Washington, DC: U.S. Government Printing Office.

Butler, Declan. 2006. "Virtual Globes: The Web-Wide World." *Nature* 439 (7078): 776–78.

Buttimer, Anne, and David Seamon, eds. 1980. *The Human Experience of Space and Place.* Kent, UK: Croom Helm.

Calvino, Italo. 1965. *Cosmicomics.* Translated by William Weaver. New York: Harcourt Brace & Company.

Cambrosio, Albert, Daniel Jacobi, and Peter Keating. 1993. "Ehrlich's 'Beautiful Pictures' and the Controversial Beginnings of Immunological Imagery." *Isis* 84 (4): 662–99.

Canales, Jimena. 2002a. "Photogenic Venus: The 'Cinematographic Turn' and Its Alternatives in Nineteenth-Century France." *Isis* 93 (4): 585–613.

Canales, Jimena. 2002b. *Photogenic Venus: The "Cinematographic Turn" and Its Alternatives in Nineteenth-Century France.* Chicago: University of Chicago Press.

Canales, Jimena. 2010. *A Tenth of a Second: A History.* Chicago: University of Chicago Press.

Casey, Edward S. 1993. *Getting Back into Place: Toward a Renewed Understanding of the Place-World.* Bloomington: Indiana University Press.

Casey, Edward S. 1996. "How to Get from Space to Place in a Fairly Short Stretch of Time: Phenomenological Prolegomena." In *Senses of Place,* edited by Steven Feld and Keith H. Basso, 13–52. Santa Fe, NM: School of American Research Press.

Casey, Edward S. 1998. *The Fate of Place: A Philosophical History.* Berkeley: University of California Press.

Catling, D. C. 2004. "Planetary Science: On Earth, as It Is on Mars?" *Nature* 429 (6993): 707–8.

Chaikin, Andrew. 2008. *A Passion for Mars: Intrepid Explorers of the Red Planet.* New York: Abrams.

Chakrabarty, Dipesh. 2009. "The Climate of History: Four Theses." *Critical Inquiry* 35 (2): 197–222.

Chamberlain, Joseph, and Dale Cruikshank. 1999. "The Beginnings of the Division for Planetary Science of the AAS." In *The American Astronomical Society's First Century,* vol. 1, edited by David H. DeVorkin, 252. Washington, DC: American Institute of Physics.

Chan, M. A., B. Beitler, W. T. Parry, J. Ormö, and G. Komatsu. 2004. "A Possible Terrestrial Analogue for Haematite Concretions on Mars." *Nature* 429 (6993): 731–34.

Charbonneau, D., Z. K. Berta, J. Irwin, C. J. Burke, P. Nutzman, L. A. Buchhave, C. Lovis, et al. 2009. "A Super-Earth Transiting a Nearby Low-Mass Star." *Nature* 462 (7275): 891–94.

Childress, April. 2003. "My Father's Birthday." Mars Desert Research Station Crew

Reports. March 24, http://web.archive.org/web/20060621130750/http://www
.marssociety.org/MDRS/fs02/0324/nar.asp.

Christiansen, J., et al. 2010. "Studying the Atmosphere of the Exoplanet Hat-P-7b via
Secondary Eclipse Measurement with EPOXI, Spitzer, and Kepler." *Astrophysical
Journal* 710:97–104.

Clancey, William. 2002. "Commander's Logbook." Mars Desert Research Station
Crew Reports. April 19, http://web.archive.org/web/20060203131744/http://www
.marssociety.org/MDRS/fs01/0419/log.asp.

Coleman, Gabriella. 2004. "The Political Agnosticism of Free and Open Source Software and the Inadvertent Politics of Contrast." *Anthropological Quarterly* 77 (3):
507–19.

Coleman, Gabriella, and Alex Golub. 2008. "Hacker Practice: Moral Genres and the
Cultural Articulation of Liberalism." *Anthropological Theory* 8 (3): 255–77.

Collins, Harry. 2004. *Gravity's Shadow: The Search for Gravitational Waves*. Chicago: University of Chicago Press.

Collins, Harry M. 1974. "The TEA Set: Tacit Knowledge and Scientific Networks." *Science Studies* 4 (2): 165–85.

Corner, James. 1999. "The Agency of Mapping: Speculation, Critique and Invention."
In *Mappings*, edited by Denis E. Cosgrove, 213–52. London: Reaktion Books.

Cosgrove, Denis E. 1985. "Prospect, Perspective and the Evolution of the Landscape
Idea." *Transactions of the Institute of British Geographers* 10 (1): 45–62.

Cosgrove, Denis E. 1994. "Contested Global Visions: One-World, Whole-Earth, and
the Apollo Space Photographs." *Annals of the Association of American Geographers* 84 (2):
270–94.

Cosgrove, Denis E. 2001. *Apollo's Eye: A Cartographic Genealogy of the Earth in the Western
Imagination*. Baltimore, MD: Johns Hopkins University Press.

Cowan, Nicolas B., Eric Agol, Victoria S. Meadows, Tyler Robinson, Timothy A.
Livengood, Drake Deming, Carey M. Lisse, et al. 2009. "Alien Maps of an Ocean-
Bearing World." *Astrophysical Journal* 700 (2): 915–23.

Crampton, Jeremy W. 2009. "Cartography: Maps 2.0." *Progress in Human Geography* 33
(1): 91–100.

Crary, Jonathan. 1990. *Techniques of the Observer: On Vision and Modernity in the Nineteenth
Century*. Cambridge, MA: MIT Press.

Cresswell, Tim. 1996. *In Place/Out of Place: Geography, Ideology, and Transgression*. Minneapolis: University of Minnesota Press.

Crowe, Michael J. 1986. *The Extraterrestrial Life Debate, 1750–1900*. Cambridge: Cambridge University Press.

Crowe, Michael J. 2008. *The Extraterrestrial Life Debate, Antiquity to 1915: A Source Book*.
South Bend, IN: University of Notre Dame Press.

Csicsery-Ronay, Istvan. 2008. *The Seven Beauties of Science Fiction*. Middletown, CT: Wes-
leyan University Press.

Dale, Ashley. 2011. "Journalist's Report." Mars Desert Research Station Crew Reports.

December 3, https://sites.google.com/a/marssociety.org/MDRS/home/field
-reports/crew108/day01.

Daston, Lorraine, ed. 2000. *Biographies of Scientific Objects*. Chicago: University of Chicago Press.

Daston, Lorraine, and Peter Galison. 2007. *Objectivity*. New York: Zone Books.

Daston, Lorraine, and Elizabeth Lunbeck. 2011. *Histories of Scientific Observation*. Chicago: University of Chicago Press.

Davidson, Keay. 1999. *Sagan: A Life*. New York: Wiley.

Davis, Morgan. 1970. "The New Geology." *Bulletin of the Geological Society of America* 81 (2): 331.

Dean, Jodi. 1998. *Aliens in America: Conspiracy Cultures from Outerspace to Cyberspace*. Ithaca, NY: Cornell University Press.

de Certeau, Michel. 1988. *The Practice of Everyday Life*. Translated by Steven Rendell. Berkeley: University of California Press.

DeGroot, Gerard. 2006. *Dark Side of the Moon: The Magnificent Madness of the American Lunar Quest*. New York: New York University Press.

Del Casino, Vincent J., and Stephen P. Hanna. 2006. "Beyond the 'Binaries': A Methodological Intervention for Interrogating Maps as Representational Practices." *ACME: An International E-Journal for Critical Geographies* 4 (1): 34–56.

Descola, Philippe. 2010. "Cognition, Perception and Worlding." *Interdisciplinary Science Reviews* 35 (3–4): 334–40.

Diamond-Lowe, H., et al. 2015. "New Analysis Indicates No Thermal Inversion in the Atmosphere of HD 209458b." *Astrophysical Journal* 796:66–72.

Dick, Steven. 1984. *Plurality of Worlds: The Origins of the Extraterrestrial Life Debate from Democritus to Kant*. Cambridge: Cambridge University Press.

Dick, Steven, and James Strick. 2004. *The Living Universe: NASA and the Development of Astrobiology*. New Brunswick, NJ: Rutgers University Press.

Doel, Ronald E. 1996. *Solar System Astronomy in America: Communities, Patronage, and Interdisciplinary Science, 1920–1960*. Vol. 1. Cambridge: Cambridge University Press.

Dole, Stephen H. 1964. *Habitable Planets for Man*. Santa Monica, CA: RAND Corporation.

Dumit, Joseph. 2004. *Picturing Personhood: Brain Scans and Biomedical Identity*. Princeton, NJ: Princeton University Press.

Duncan, James. 2005. *The City as Text: The Politics of Landscape Interpretation in the Kandyan Kingdom*. Cambridge: Cambridge University Press.

Duncan, James, and Nancy Duncan. 1988. "(Re)Reading the Landscape." *Environment and Planning D: Society and Space* 6 (2): 117–26.

Dutton, Geoffrey. 1996. "Encoding and Handling Geospatial Data with Hierarchical Triangular Meshes." In *Advances in GIS Research II*, edited by M. Kraak and M. Molenaar, 505–18. London: Taylor & Francis.

Edge, David, and Michael Mulkay. 1976. *Astronomy Transformed: The Emergence of Radio Astronomy in Britain*. New York: Wiley.

Edmondson, Frank Kelly. 1997. *AURA and Its US National Observatories*. Cambridge: Cambridge University Press.

Edney, Matthew H. 1997. *Mapping an Empire: The Geographical Construction of British India, 1765–1843*. Vol. 10. Chicago: University of Chicago Press.

Elsaesser, Thomas. 2010. "The Dimension of Depth and Objects Rushing towards Us. Or: The Tail That Wags the Dog. A Discourse on Digital 3-D Cinema." *eDIT Filmmaker's Magazine*, vol 1.

Elsaesser, Thomas. 2013. "The 'Return' of 3-D: On Some of the Logics and Genealogies of the Image in the Twenty-First Century." *Critical Inquiry* 39 (2): 217–46.

Entrikin, J. Nicholas. 1991. *The Betweenness of Place: Towards a Geography of Modernity*. Baltimore, MD: Johns Hopkins University Press.

Fabian, Johannes. 1983. *Time and the Other: How Anthropology Constructs Its Object*. New York: Columbia University Press.

Farman, Jason. 2010. "Mapping the Digital Empire: Google Earth and the Process of Postmodern Cartography." *New Media and Society* 12 (6): 869–88.

Ferris, Timothy. 2009. "Worlds Apart: Seeking New Earths." *National Geographic*.

Fontenelle, Bernard le Bovier de. 1990. *Conversations on the Plurality of Worlds*. Translated by H. A. Hargreaves. Berkeley: University of California Press.

Foucault, Michel. 1967. "Of Other Spaces." Translated by Jay Miskowiec. *Diacritics* 16:22–27.

Fox, William L. 2006. *Driving to Mars*. Emeryville, CA: Shoemaker and Hoard.

Friedman, Susan Stanford. 2010. "Planetarity: Musing Modernist Studies." *Modernism/modernity* 17 (3): 471–99.

Frischauf, Norbert. 2006. "Commander's Journal." Mars Desert Research Station Crew Reports. April 20, http://web.archive.org/web/20060927090425/http://www.marssociety.org/MDRS/fs05/0420/cmdr.asp.

Fuller, David. 2003. "Commander's Report." Mars Desert Research Station Crew Reports. March 26, http://web.archive.org/web/20060621074435/http://www.marssociety.org/MDRS/fs02/0326/cmdr.asp.

Gale, Celeste. 2005a. "Commander's Log." Mars Desert Research Station Crew Reports. January 13, http://web.archive.org/web/20060110152509/http://www.marssociety.org/MDRS/fs04/0113/cmdr.asp.

Gale, Celeste. 2005b. "Commander's Report." Mars Desert Research Station Crew Reports. January 10, http://web.archive.org/web/20051119130143/http://www.marssociety.org/MDRS/fs04/0110/cmdr.asp.

Galison, Peter. 1997. *Image and Logic: A Material Culture of Microphysics*. Chicago: University of Chicago Press.

Galison, Peter. 2003. "The Collective Author." In *Scientific Authorship: Credit and Intellectual Property in Science*, edited by Mario Biagioli and Peter Galison, 325–55. New York: Routledge.

Galison, Peter, and Emily Thompson, eds. 1999. *The Architecture of Science*. Cambridge, MA: MIT Press.

Garb, Y. J. 1985. "The Use and Misuse of the Whole Earth Image." *Whole Earth Review*, March, 18–25.

Gartner, Georg. 2009. "Web Mapping 2.0." In *Rethinking Maps: New Frontiers in Cartographic Theory*, edited by Martin Dodge, Rob Kitchin, and Chris Perkins, 68–82. New York: Routledge.

Geertz, Clifford. 1988. *Works and Lives: The Anthropologist as Author.* Palo Alto, CA: Stanford University Press.

Gentner, Dedre, and Michael Jeziorski. 1993. "The Shift from Metaphor to Analogy in Western Science." In *Metaphor and Thought*, edited by Andrew Ortony, 447–80. Cambridge: Cambridge University Press.

"Geographical Notes." 1886. *Proceedings of the Royal Geographical Society and Monthly Record of Geography* 8 (1): 47–52.

Gieryn, Thomas. 2002. "Three Truth-Spots." *Journal of the History of the Behavioral Sciences* 38 (2): 113–32.

Gieryn, Thomas. 2006. "City as Truth-Spot: Laboratories and Field-Sites in Urban Studies." *Social Studies of Science* 36 (1): 5–38.

Gilroy, Paul. 2005. *Postcolonial Melancholia.* New York: Columbia University Press.

Goodwin, Charles. 1995. "Seeing in Depth." *Social Studies of Science* 25 (2): 237–74.

Gough, Mike. 2002. "Commander's Report." Mars Desert Research Station Crew Reports. December 9, http://web.archive.org/web/20060621115747/http://www.marssociety.org/MDRS/fs02/1209/cmdr.asp.

Grasseni, Cristina, ed. 2009. *Skilled Visions: Between Apprenticeship and Standards.* New York: Berghahn Books.

Greeley, Ronald, and Raymond M. Batson, eds. 1990. *Planetary Mapping.* Cambridge: Cambridge University Press.

Green, J., and D. Wolfle. 1960. "Gea, Daughter of Chaos." *Science* 131 (3407): 1071.

Griffiths, Alison. 2008. *Shivers down Your Spine: Cinema, Museums, and the Immersive View.* New York: Columbia University Press.

Gupta, Akhil, and James Ferguson. 1992. "Beyond 'Culture': Space, Identity, and the Politics of Difference." *Cultural Anthropology* 7 (1): 6–23.

Haklay, Mordechai. 2013. "Neogeography and the Delusion of Democratisation." *Environment and Planning A* 45 (1): 55–69.

Hallyn, Fernand, ed. 2000. *Metaphor and Analogy in the Sciences.* Dordrecht, NL: Kluwer Academic.

Hampton, Keith N., Lauren Sessions Goulet, and Garrett Albanesius. 2015. "Change in the Social Life of Urban Public Spaces: The Rise of Mobile Phones and Women, and the Decline of Aloneness over 30 Years." *Urban Studies* 52 (8): 1489–504.

Haraway, Donna. 1988. "Situated Knowledges: The Science Question in Feminism and the Privilege of Partial Perspective." *Feminist Studies* 14 (3): 575–99.

Haraway, Donna. 1990. *Primate Visions: Gender, Race, and Nature in the World of Modern Science.* New York: Routledge.

Harley, J. B. 1989. "Deconstructing the Map." *Cartographica* 26 (2): 1–20.

Hartman, Edwin. 1970. *Adventures in Research: A History of Ames Research Center, 1940–1965.* SP-4302. Washington, DC: NASA.

Harvey, David. 1989. *The Condition of Postmodernity: An Enquiry into the Origins of Cultural Change.* Cambridge: Blackwell.

Harvey, David. 1993. "From Space to Place and Back Again: Reflections on the Condition of Postmodernity." In *Mapping the Futures: Local Cultures, Global Change*, edited by Jon Bird, Barry Curtis, Tim Putnam, George Robertson, and Lisa Tickner, 2–29. London: Routledge.

Hastrup, Kirsten. 1995. *A Passage to Anthropology: Between Experience and Theory.* London: Routledge.

Hayden, Corinne. 2003. *When Nature Goes Public: The Making and Unmaking of Bioprospecting in Mexico.* Princeton, NJ: Princeton University Press.

Hearnshaw, J. 1997. "Photometry, Astronomical." In *History of Astronomy: An Encyclopedia*, edited by John Lankford, 395–401. New York: Garland.

Hegazy, Ashraf. 2004. "Crew Narrative." Mars Desert Research Station Crew Reports. January 9, http://web.archive.org/web/20051225225943/http://www.marssociety.org/MDRS/fs03/0107/nar.asp.

Heidegger, Martin. 1927. *Being and Time.* Translated by John Macquarrie and Edward Robinson. San Francisco, CA: Harper.

Heidegger, Martin. 1951. "Building Dwelling Thinking." In *Poetry, Language, Thought*, translated by Albert Hofstadter, 141–60. New York: Harper Collins.

Heidegger, Martin. 1966. "Only a God Can Save Us: The Spiegel Interview." In *Heidegger: The Man and the Thinker*, edited by Thomas Sheehan, translated by William Richardson, 45–68. New Brunswick, NJ: Transaction.

Helmreich, Stefan. 2000. *Silicon Second Nature: Culturing Artificial Life in a Digital World.* Berkeley: University of California Press.

Helmreich, Stefan. 2007. "An Anthropologist Underwater: Immersive Soundscapes, Submarine Cyborgs, and Transductive Ethnography." *American Ethnologist* 34 (4): 621–41.

Helmreich, Stefan. 2009. *Alien Ocean: Anthropological Voyages in Microbial Seas.* Berkeley: University of California Press.

Henke, Christopher. 2000. "Making a Place for Science: The Field Trial." *Social Studies of Science* 30 (4): 483–511.

Hesse, Mary B. 1963. *Models and Analogies in Science.* London: Sheed and Ward.

Hirsch, E. 1995. "Introduction: Landscape: Between Place and Space." In *The Anthropology of Landscape: Perspectives on Place and Space*, edited by E. Hirsch and M. O'Hanlon, 1–30. Oxford: Clarendon Press.

Hoeppe, G. 2012. "Astronomers at the Observatory: Place, Visual Practice, Traces." *Anthropological Quarterly* 85 (4): 1141–60.

Holton, Gerald. 1998. *The Scientific Imagination.* Cambridge, MA: Harvard University Press.

hooks, bell. 1990. "Homeplace: A Site of Resistance." In *Yearning: Race, Gender, and Cultural Politics*, 41–50. Boston, MA: South End Press.

Huang, Su-Shu. 1959. "Occurrence of Life in the Universe." *American Scientist* 47 (3): 397–402.

Hubbert, M. K. 1963. "Are We Retrogressing in Science?" *Geological Society of America Bulletin* 74:365–78.

"Improvements in Photometry." 1895. *Nature* 51 (1328): 558–61.

Ingold, Tim. 1993. "Globes and Spheres." In *Environmentalism: The View from Anthropology*, edited by Kay Milton, 31–42. New York: Routledge.

Jameson, Fredric. 1982. "Progress versus Utopia; Or, Can We Imagine the Future?" *Science Fiction Studies* 9 (2): 147–58.

Jameson, Fredric. 2005. "'If I Can Find One Good City, I Will Spare the Man': Realism and Utopia in Kim Stanley Robinson's Mars Trilogy." In *Archaeologies of the Future: The Desire Called Utopia and Other Science Fictions*, 393–416. New York: Verso Books.

Janesick, J., and T. Elliott. 1992. "History and Advancement of Large Array Scientific CCD Imagers." In *Astronomical CCD Observing and Reduction Techniques*, 23:1–67.

Jasanoff, Sheila. 2004. "Heaven and Earth: The Politics of Environmental Images." In *Earthly Politics: Local and Global in Environmental Governance*, edited by Sheila Jasanoff and Marybeth Long Martello, 31–52. Cambridge, MA: MIT Press.

Jazeel, Tariq. 2011. "Spatializing Difference beyond Cosmopolitanism: Rethinking Planetary Futures." *Theory, Culture and Society* 28 (5): 75–97.

Joyce, K. 2005. "Appealing Images: Magnetic Resonance Imaging and the Production of Authoritative Knowledge." *Social Studies of Science* 35 (3): 437–62.

Kaiser, David. 2000. "Stick-Figure Realism: Conventions, Reification, and the Persistence of Feynman Diagrams, 1948–1964." *Representations* 70:49–86.

Kaiser, David. 2005a. *Drawing Theories Apart: The Dispersion of Feynman Diagrams in Postwar Physics*. Chicago: University of Chicago Press.

Kaiser, David. 2005b. *Pedagogy and the Practice of Science: Historical and Contemporary Perspectives*. Cambridge, MA: MIT Press.

Kasting, James F., Daniel P. Whitmire, and Ray T. Reynolds. 1993. "Habitable Zones around Main Sequence Stars." *Icarus* 101 (1): 108–28.

Kay, Lily E. 2000. *Who Wrote the Book of Life? A History of the Genetic Code*. Stanford, CA: Stanford University Press.

Keim, Brandon. 2009. "Most Earth-Like Extrasolar Planet Found Right Next Door." *Wired Science: News for Your Neurons*. December 16, http://www.wired.com/wired science/2009/12/super-earth/.

Kelty, Christopher. 2008. *Two Bits: The Cultural Significance of Free Software*. Durham, NC: Duke University Press.

Kemp, Martin. 1997. "Seeing and Picturing: Visual Representation in Twentieth-Century Science." In *Science in the Twentieth Century*, edited by John Krige and Dominique Pestre, 361–90. Amsterdam, NL: Harwood Academic.

Kennefick, Daniel. 2000. "Star Crushing: Theoretical Practice and the Theoreticians' Regress." *Social Studies of Science* 30 (1): 5–40.

Kern, Stephen. 2003. *The Culture of Time and Space, 1880–1918*. Cambridge, MA: Harvard University Press.

Kitchin, Rob, Justin Gleeson, and Martin Dodge. 2013. "Unfolding Mapping Practices: A New Epistemology for Cartography." *Transactions of the Institute of British Geographers* 38 (3): 480–96.

Knorr-Cetina, Karin. 1999. *Epistemic Cultures: How the Sciences Make Knowledge*. Cambridge, MA: Harvard University Press.

Knutson, Heather A., David Charbonneau, Lori E. Allen, Adam Burrows, and S. Thomas Megeath. 2008. "The 3.6–8.0 μm Broadband Emission Spectrum of HD 209458b: Evidence for an Atmospheric Temperature Inversion." *Astrophysical Journal* 673 (1): 526–31.

Kohler, Robert E. 2002. *Landscapes and Labscapes: Exploring the Lab-Field Border in Biology*. Chicago: University of Chicago Press.

Kopal, Zdenek, and A. G. Wilson. 1962. "Preface." *Icarus* 1 (1): i–iii.

Kuklick, Henrika, and Robert Kohler. 1996. "Introduction to Special Issue on Science in the Field." *Osiris* 11:1–14.

Lakoff, George, and Mark Johnson. 1980. *Metaphors We Live By*. Chicago: University of Chicago Press.

Lane, K. Maria D. 2009. "Astronomers at Altitude: Mountain Geography and the Cultivation of Scientific Legitimacy." In *High Places: Cultural Geographies of Mountains, Ice, and Science*, edited by Denis Cosgrove and Veronica della Dora, 126–44. London: I. B. Tauris.

Lane, K. Maria D. 2010. *Geographies of Mars: Seeing and Knowing the Red Planet*. Chicago: University of Chicago Press.

Lankford, John. 1997. *American Astronomy: Community, Careers, and Power, 1859–1940*. Chicago: University of Chicago Press.

Latour, Bruno. 1987. *Science in Action: How to Follow Scientists and Engineers through Society*. Cambridge, MA: Harvard University Press.

Latour, Bruno. 1990. "Drawing Things Together." In *Representation in Scientific Practice*, edited by Michael Lynch and Steve Woolgar, 19–68.Cambridge, MA: MIT Press.

Latour, Bruno. 2014. "Agency at the Time of the Anthropocene." *New Literary History* 45 (1): 1–18.

Latour, Bruno. 2015. "Telling Friends from Foes at the Time of the Anthropocene." In *The Anthropocene and the Global Environmental Crisis: Rethinking Modernity in a New Epoch*, edited by Clive Hamilton, Christophe Bonneuil, and François Gemenne, 145–54. New York: Routledge.

Latour, Bruno, and Steve Woolgar. 1986. *Laboratory Life: The Construction of Scientific Facts*. Princeton, NJ: Princeton University Press.

Launius, Roger D. 2013. *Exploring the Solar System: The History and Science of Planetary Exploration*. New York: Palgrave Macmillan.

Lawson, P. R., W. A. Traub, and S.C. Unwin, eds. 2009. "Exoplanet Community Report." Washington, DC: NASA.

Lefebvre, Henri. 1974. *The Production of Space*. Translated by Donald Nicholson-Smith. Oxford: Blackwell.

Lemke, Jay. 1998. "Multiplying Meaning: Visual and Verbal Semiotics in Scientific Text." In *Reading Science: Critical and Functional Perspectives on Discourses of Science*, edited by J. R. Martin and Robert Veel, 87–113. London: Routledge.

Lemonick, Michael D. 1998. *Other Worlds: The Search for Life in the Universe*. New York: Simon and Schuster.

Lempert, William. 2014. "Decolonizing Encounters of the Third Kind: Alternative Futuring in Native Science Fiction Film." *Visual Anthropology Review* 30 (2): 164–76.

Lepselter, Susan. 1997. "From the Earth Native's Point of View: The Earth, the Extraterrestrial and the Natural Ground of Home." *Public Culture* 9:197–208.

Limerick, Patricia. 1988. *The Legacy of Conquest: The Unbroken Past of the American West*. New York: Norton.

Limerick, Patricia. 1992. "Imagined Frontiers: Westward Expansion and the Future of the Space Program." In *Space Policy Alternatives*, edited by Radford Byerly, 249–62. Boulder, CO: Westview Press.

Livingstone, David. 2003. *Putting Science in Its Place: Geographies of Scientific Knowledge*. Chicago: University of Chicago Press.

Logsdon, John M. 1970. *The Decision to Go to the Moon: Project Apollo and the National Interest*. Cambridge, MA: MIT Press.

Lovelock, James. 1979. *Gaia: A New Look at Life on Earth*. Oxford: Oxford University Press.

Low, Setha. 2009. "Towards an Anthropological Theory of Space and Place." *Semiotica* 175 (1/4): 21–37.

Low, Setha, and Denise Lawrence-Zúñiga. 2003. *The Anthropology of Space and Place: Locating Culture*. Malden, MA: Blackwell.

Lowe, Celia. 2006. *Wild Profusion: Biodiversity Conservation in an Indonesian Archipelago*. Princeton, NJ: Princeton University Press.

Lunine, Jonathan. 2004. "Jonathan Lunine: Life Finder." *Astrobiology Magazine*. May 9, http://www.astrobio.net/topic/exploration/moon-to-mars/jonathan-lunine-life-finder/.

Lunine, Jonathan, Debra Fischer, Heidi Hammel, Thomas Henning, Lynne Hillenbrand, James Kasting, Greg Laughlin, et al. 2008. *Worlds Beyond: A Strategy for Detection and Characterization of Exoplanets*. Report of the ExoPlanet Task Force. Washington, DC: Astronomy and Astrophysics Advisory Committee.

Lynch, Michael. 1985. *Art and Artifact in Laboratory Science: A Study of Shop Work and Shop Talk in a Research Laboratory*. London: Routledge and Kegan Paul.

Lynch, Michael. 1991a. "Laboratory Space and the Technological Complex: An Investigation of Topical Contextures." *Science in Context* 4 (1): 51–78.

Lynch, Michael. 1991b. "Science in the Age of Mechanical Reproduction: Moral and

Epistemic Relations between Diagrams and Photographs." *Biology and Philosophy* 6 (2): 205–26.

Lynch, Michael, and Samuel Edgerton. 1988. "Aesthetics and Digital Image Processing: Representational Craft in Contemporary Astronomy." In *Picturing Power: Visual Depiction and Social Relations*, edited by Gordon Fyfe and John Law, 184–220. London: Routledge.

Lynch, Michael, and Steve Woolgar. 1990. *Representation in Scientific Practice*. Cambridge, MA: MIT Press.

Mack, Pamela Etter. 1990. *Viewing the Earth: The Social Construction of the Landsat Satellite System*. Cambridge, MA: MIT Press.

Madhusudhan, N., and S. Seager. 2009. "A Temperature and Abundance Retrieval Method for Exoplanet Atmospheres." *Astrophysical Journal* 707 (1): 24–39.

Mannheim, Karl. 1929. *Ideology and Utopia: An Introduction to the Sociology of Knowledge*. Translated by Louis Wirth and Edward Shils. New York: Harcourt Trade.

Marcus, George E. 1995. "Ethnography in/of the World System: The Emergence of Multi-sited Ethnography." *Annual Review of Anthropology* 24 (1): 95–117.

Marcus, George E., and Michael M. J. Fischer. 1999. *Anthropology as Cultural Critique: An Experimental Moment in the Human Sciences*. 2nd ed. Chicago: University of Chicago Press.

Markley, Robert. 2005. *Dying Planet: Mars in Science and the Imagination*. Durham, NC: Duke University Press.

Marx, Leo. 1964. *The Machine in the Garden: Technology and the Pastoral Ideal in America*. Oxford: Oxford University Press.

Masco, Joseph. 2010. "Bad Weather: On Planetary Crisis." *Social Studies of Science* 40 (1): 7–40.

Massey, Doreen. 1994a. "A Place Called Home?" In *Space, Place, and Gender*, 157–74. Minneapolis: University of Minnesota Press.

Massey, Doreen. 1994b. *Space, Place, and Gender*. Minneapolis: University of Minnesota Press.

Massey, Doreen. 2005. "The Elusiveness of Place." In *For Space*, 130–42. London: Sage.

Mayor, M., and D. Queloz. 1995. "A Jupiter-Mass Companion to a Solar-Type Star." *Nature* 378 (6555): 355–59.

McCray, Patrick. 2004. *Giant Telescopes: Astronomical Ambition and the Promise of Technology*. Cambridge, MA: Harvard University Press.

McDougall, Walter A. 1985. *The Heavens and the Earth: A Political History of the Space Age*. New York: Basic Books.

Merrifield, Andrew. 1993. "Place and Space: A Lefebvrian Reconciliation." *Transactions of the Institute of British Geographers* 18 (4): 516–31.

Messeri, Lisa. 2010. "The Problem with Pluto." *Social Studies of Science* 40 (2): 187–214.

Messeri, Lisa, Carol Stoker, and Bernard Foing. 2010. "How Will Astronauts Spend Their Time on the Moon? Insights from a Field Mission at the Mars Desert Research Station, Utah." Poster presented at the NASA Lunar Science Institute, July, Mountain View, CA.

Messeri, Lisa, and Janet Vertesi. 2015. "The Greatest Missions Never Flown: Anticipatory Discourse and the Projectory in Technological Communities." *Technology and Culture* 56 (1): 54–85.

Mindell, David A. 2008. *Digital Apollo: Human, Machine, and Space Flight*. Cambridge, MA: MIT Press.

Mirmalek, Zara. 2008. "Solar Discrepancies: Mars Exploration and the Curious Problem of Interplanetary Time." Ph.D. diss., University of California, San Diego.

Mitchell, Don. 1996. *The Lie of the Land: Migrant Workers and the California Landscape*. Minneapolis: University of Minnesota Press.

Mitchell, Don. 2005. "Landscape." In *Cultural Geography: A Critical Dictionary of Key Concepts*, edited by David Atkinson, Peter Jackson, David Sibley, and Neil Washbourne, 49–56. New York: I.B. Tauris.

Mol, Annemarie. 2002. *The Body Multiple: Ontology in Medical Practice*. Durham, NC: Duke University Press.

Moran, Patrick. 2003. "Developing an Open Source Option for NASA Software." NAS Technical report NAS-03-009. NAS Technical Report. Moffett Field: NASA Ames Research Center.

Morgan-Dimmick, Pete. 2014. "Journalist's Report." Mars Desert Research Station Crew Reports. May 6, http://MDRS.marssociety.org/home/crew-141/506 -journalistreport.

Morgan, Robin. 1984. "Planetary Feminism: The Politics of the 21st Century." In *Sisterhood Is Global: The International Women's Movement Anthology*, edited by Robin Morgan, 1–37. New York: Feminist Press at the City University of New York.

Moy, Timothy. 2004. "Culture, Technology, and the Cult of Tech in the 1970s." In *America in the 70s*, edited by Beth Bailey and David Farber, 208–28. Lawrence: University of Kansas Press.

Moylan, Tom. 1986. *Demand the Impossible: Science Fiction and the Utopian Imagination*. New York: Methuen.

Munns, David P. D. 2012. *A Single Sky: How an International Community Forged the Science of Radio Astronomy*. Cambridge, MA: MIT Press.

Mutch, Thomas. 1978. *The Martian Landscape*. SP-425. Washington, DC: NASA.

Myers, F. 1993. "Place, Identity, and Exchange in a Totemic System: Nurturance and the Process of Social Reproduction in Pintupi Society." In *Exchanging Products, Producing Exchange*, edited by J. Fajans, 33–57. Sydney, AU: University of Sydney.

Myers, F. R. 2000. "Ways of Place-Making." In *Culture, Landscape, and the Environment: The Linacre Lectures 1997*, edited by Kate Flint and Howard Morphy, 101–19. Oxford: Oxford University Press.

Myers, Fred. 1991. *Pintupi Country, Pintupi Self: Sentiment, Place, and Politics among Western Desert Aborigines*. Berkeley: University of California Press.

Nicholson, Jean M. 1979. "Mars—Complex and Mysterious." *Altadena (CA) Chronicle*, January 19.

Nisbett, Catherine. 2007. "Business Practice: The Rise of American Astrophysics, 1859–1919." PhD diss., Princeton University, Princeton, NJ.

Nye, David E. 1997. *Narratives and Spaces: Technology and the Construction of American Culture.* New York: Columbia University Press.

Nye, David E. 2003. *America as Second Creation: Technology and Narratives of New Beginnings.* Cambridge, MA: MIT Press.

Olson, Valerie. 2010. "American Extreme: An Ethnography of Astronautical Visions and Ecologies." PhD diss., Rice University, Houston, TX.

Olson, Valerie, and Lisa Messeri. 2015. "Beyond the Anthropocene: Un-Earthing an Epoch." *Environment and Society: Advances in Research* 6:28–47.

Overbye, Dennis. 2009a. "In a Lonely Cosmos, a Hunt for Worlds Like Ours." *New York Times*, March 3, Science sec. http://www.nytimes.com/2009/03/03/science/03kepl .html?_r=1&scp=7&sq=kepler%201aunch&st=cse.

Overbye, Dennis. 2009b. "A Sultry World Is Found Orbiting a Distant Star." *New York Times*, December 17, Science sec., Space and Cosmos. http://www.nytimes.com /2009/12/17/science/space/17planet.html?_r=1&scp=1&sq=overbye+sultry+world &st=nyt.

Pallé, E., E. B. Ford, S. Seager, P. Montañés-Rodríguez, and M. Vázquez. 2008. "Identifying the Rotation Rate and the Presence of Dynamic Weather on Extrasolar Earthlike Planets from Photometric Observations." *Astrophysical Journal* 676:1319–29.

Pang, Alex Soojung-Kim. 1996. "Gender, Culture, and Astrophysical Fieldwork: Elizabeth Campbell and the Lick Observatory-Crocker Eclipse Expeditions." *Osiris* 11:17–43.

Pang, Alex Soojung-Kim. 2002. *Empire and the Sun: Victorian Solar Eclipse Expeditions.* Stanford, CA: Stanford University Press.

Parker, Martin. 2009. "Capitalists in Space." *Sociological Review* 57 (1): 83–97.

Pickles, John. 2004. *A History of Spaces: Cartographic Reason, Mapping and the Geo-Coded World.* New York: Routledge.

Pina-Cabral, João de. 2014. "World: An Anthropological Examination (Part 1)." HAU: *Journal of Ethnographic Theory* 4 (1): 49–73.

Pinch, Trevor J. 1986. *Confronting Nature: The Sociology of Solar-Neutrino Detection.* Dordrecht: D. Reidel.

Plavchan, P. 2010. "Introduction to Time-Series and Identifying Variability and Periodicities." Paper presented at the 2010 Sagan Exoplanet Summer Workshop, "Stars as Homes for Habitable Planetary Systems," California Institute of Technology, July 26.

Polanyi, Michael. 1958. *Personal Knowledge.* London: Routledge.

Poole, Robert. 2010. *Earthrise: How Man First Saw the Earth.* New Haven, CT: Yale University Press.

Powell, Richard C. 2007. "Geographies of Science: Histories, Localities, Practices, Futures." *Progress in Human Geography* 31 (3): 309–29.

Pratt, Mary Louise. 1992. *Imperial Eyes: Travel Writing and Transculturation*. New York: Routledge.

Pravdo, Steven, and Stuart Shaklan. 2009. "An Ultracool Star's Candidate Planet." *Astrophysical Journal* 700:623.

Pred, Allan. 1984. "Place as Historically Contingent Process: Structuration and the Time-Geography of Becoming Places." *Annals of the Association of American Geographers* 74 (2): 279–97.

Putnam, Robert D. 2000. *Bowling Alone: The Collapse and Revival of American Community*. New York: Simon and Schuster.

Queloz, D., F. Bouchy, C. Moutou, A. Hatzes, G. Hébrard, R. Alonso, M. Auvergne, et al. 2009. "The CoRoT-7 Planetary System: Two Orbiting Super-Earths." *Astronomy and Astrophysics* 506 (1): 303–19.

Raffles, Hugh. 2002. *In Amazonia: A Natural History*. Princeton, NJ: Princeton University Press.

Rajpaul, Vinesh, Suzanne Aigrain, and S. Roberts. 2016. "Ghost in the time series: no planet for Alpha Cen B." *Monthly Notices of the Royal Astronomical Society: Letters* 456 (1): L6–10.

Rand, Lisa Ruth. 2014. "The Case of Female Astronauts: Reproducing Americans in the Final Frontier." *Appendix* 2 (3). http://theappendix.net/issues/2014/7/the-case -for-female-astronauts-reproducing-americans-in-the-final-frontier.

Rankama, K. 1962. "Planetology and Geology." *Bulletin of the Geological Society of America* 73 (4): 519–20.

Redfield, Peter. 2000. *Space in the Tropics: From Convicts to Rockets in French Guiana*. Berkeley: University of California Press.

Redfield, Peter. 2002. "The Half-Life of Empire in Outer Space." *Social Studies of Science* 32 (5/6): 791–825.

Relph, Edward. 1976. *Place and Placelessness*. London: Pion.

Rheinberger, Hans-Jorg. 1997. *Toward a History of Epistemic Things: Synthesizing Proteins in the Test Tube*. Stanford, CA: Stanford University Press.

Robertson, Paul, Suvrath Mahadevan, Michael Endl, and Arpita Roy. 2014. "Stellar Activity Masquerading as Planets in the Habitable Zone of the M Dwarf Gliese 581." *Science* 345(6195): 440–44.

Robinson, K. Stanley. 1993. *Red Mars*. New York: Bantam.

Ronca, L. B. 1965. "Selenology vs. Geology of the Moon Etc." *GeoTimes* 9:13.

Rose, Gillian. 1993. *Feminism and Geography: The Limits of Geographical Knowledge*. Minneapolis: University of Minnesota Press.

Rudwick, Martin. 1976. "The Emergence of a Visual Language for Geological Science." *History of Science* 14:149–95.

Ruff, Mark. 2012. "Journalist Report." Mars Desert Research Station Crew Reports.

January 1, https://sites.google.com/a/marssociety.org/MDRS/home/field-reports/crew110a/day02.

Sack, Robert David. 1997. *Homo Geographicus: A Framework for Action, Awareness, and Moral Concern.* Baltimore: Johns Hopkins University Press.

Sagan, Carl. 1994. *Pale Blue Dot: A Vision of the Human Future in Space.* New York: Random House.

Saunders, Barry F. 2009. *CT Suite: The Work of Diagnosis in the Age of Noninvasive Cutting.* Durham, NC: Duke University Press.

Schaffer, Simon. 1988. "Astronomers Mark Time: Discipline and the Personal Equation." *Science in Context* 2 (1): 115–45.

Schaffer, Simon. 2010a. "Keeping the Books at Paramatta Observatory." In *The Heavens on Earth: Observatory Techniques in the Nineteenth Century,* edited by David Aubin, Charlotte Bigg, and H. Otto Sibum, 118–47. Durham, NC: Duke University Press.

Schaffer, Simon. 2010b. "A Pattern Science." Lecture in the Tarner Lectures series, Cambridge, England. February 16, https://sms.cam.ac.uk/collection/741056.

Scheiner, J. 1894. *A Treatise on Astronomical Spectroscopy.* Translated, revised, and enlarged by Edwin Brant Frost. Boston: Ginn.

Seager, Sara. 2010. *Exoplanet Atmospheres: Physical Processes.* Princeton, NJ: Princeton University Press.

Seager, Sara. 2013. "So Many Exoplanets . . . So Few Women Scientists." *Huffington Post,* March 16. http://www.huffingtonpost.com/sara-seager/women-in-science_b_2471980.html.

Seager, Sara, and Jack Lissauer. 2010. Introduction to *Exoplanets,* edited by Sara Seager, 1. Tucson: University of Arizona Press.

Secord, James A. 2004. "Knowledge in Transit." *Isis* 95 (4): 654–72.

Seligman, C. G. 1917. "The Physical Characters of the Arabs." *Journal of the Royal Anthropological Institute of Great Britain and Ireland* 47 (June): 214–37.

Shapin, Steven. 1989. "The Invisible Technician." *American Scientist* 77 (6): 554–63.

Shapin, Steven. 1991. "'The Mind Is Its Own Place': Science and Solitude in Seventeenth-Century England." *Science in Context* 4 (01): 191–218.

Shapin, Steven. 1998. "Placing the View from Nowhere: Historical and Sociological Problems in the Location of Science." *Transactions of the Institute of British Geographers* 23(1): 5–12.

Shindell, Matthew Benjamin. 2010. "Domesticating the Planets: Instruments and Practices in the Development of Planetary Geology." *Spontaneous Generations: A Journal for the History and Philosophy of Science* 4 (1): 191–230.

Shoemaker, Eugene. 1969. "Space—Where Now, and Why?" *Engineering and Science* 33 (1): 9–12.

Siddiqi, Asif A. 2000. *Challenge to Apollo: The Soviet Union and the Space Race, 1945–1974.* NASA History Series. Washington, DC: NASA.

Silbey, Susan, and Patricia Ewick. 2003. "The Architecture of Authority: The Place of

Law in the Space of Science." In *The Place of Law*, edited by Austin Sarat, Lawrence Douglas, and Martha Umphrey, 75–108. Ann Arbor: University of Michigan Press.

Smith, J. V., and I. M. Steele. 1973. "How the Apollo Program Changed the Geology of the Moon." *Bulletin of the Atomic Scientists* 29 (9): 11–15.

Soja, Edward W. 1989. *Postmodern Geographies: The Reassertion of Space in Critical Social Theory*. London: Verso.

Soja, Edward W. 1996. *Thirdspace: Journeys to Los Angeles and Other Real-and-Imagined Places*. Cambridge: Blackwell.

Spivak, Gayatri Chakravorty. 2003. *Death of a Discipline*. New York: Columbia University Press.

Squyres, Steve. 2005. *Roving Mars: Spirit, Opportunity, and the Exploration of the Red Planet*. New York: Hyperion.

Stanley, Matthew. 2007. *Practical Mystic: Religion, Science, and A. S. Eddington*. Chicago: University of Chicago Press.

Stewart, Kathleen. 1996. *A Space on the Side of the Road: Cultural Poetics in an "Other" America*. Princeton, NJ: Princeton University Press.

Stoker, C. R., A. Zent, D. C. Catling, S. Douglas, J. R. Marshall, D. Archer Jr., B. Clark, et al. 2010. "Habitability of the Phoenix Landing Site." *Journal of Geophysical Research* 115 (E6). http://onlinelibrary.wiley.com/doi/10.1029/2009JE003421/full.

Swain, M. R., G. Vasisht, and G. Tinetti. 2008. "The Presence of Methane in the Atmosphere of an Extrasolar Planet." *Nature* 452 (7185): 329–31.

Swain, M. R., G. Vasisht, G. Tinetti, J. Bouwman, P. Chen, Y. Yung, D. Deming, et al. 2009. "Molecular Signatures in the Near-Infrared Dayside Spectrum of HD 189733b." *Astrophysical Journal Letters* 690: L114–17.

Taussig, Michael T. 1993. *Mimesis and Alterity: A Particular History of the Senses*. New York: Routledge.

Traweek, Sharon. 1988. *Beamtimes and Lifetimes: The World of High Energy Physicists*. Cambridge, MA: Harvard University Press.

Tsing, Anna. 2010. "Worlding the Matsutake Diaspora: Or, Can Actor–Network Theory Experiment with Holism?" In *Experiments in Holism: Theory and Practice in Contemporary Anthropology*, edited by Ton Otto and Nils Bubandt, 47–66. Oxford: Blackwell.

Tuan, Yi-Fu. 1977. *Space and Place: The Perspective of Experience*. Minneapolis: University of Minnesota Press.

Tuan, Yi-Fu. 1991. "Language and the Making of Place: A Narrative-Descriptive Approach." *Annals of the Association of American Geographers* 81 (4): 684–96.

Turkle, Sherry. 2012. *Alone Together: Why We Expect More from Technology and Less from Each Other*. New York: Basic Books.

Turner, Fred. 2006. *From Counterculture to Cyberculture: Stewart Brand, the Whole Earth Network, and the Rise of Digital Utopianism*. Chicago: University of Chicago Press.

Turner, Frederick Jackson. 1893. "The Significance of the Frontier in American History (1893)." In *Rereading Frederick Jackson Turner: "The Significance of the Frontier in American History," and Other Essays*, edited by John Mack Faragher, 31–60. New Haven, CT: Yale University Press.

Vakoch, Douglas A. 1998. "Signs of Life beyond Earth: A Semiotic Analysis of Interstellar Messages." *Leonardo* 31 (4): 313–19.

Valentine, David. 2012. "Exit Strategy: Profit, Cosmology, and the Future of Humans in Space." *Anthropological Quarterly* 85 (4): 1045–67.

Valentine, David, Valerie Olson, and Debbora Battaglia. 2009. "Encountering the Future: Anthropology and Outer Space." *Anthropology News* 50 (9): 11–15.

Valentine, David, Valerie Olson, and Debbora Battaglia. 2012. "Extreme: Limits and Horizons in the Once and Future Cosmos." *Anthropological Quarterly* 85 (4): 1007–26.

Vance, Ashlee. 2010. "Google and Mountain View Recast Company-Town Model." *New York Times*, February 18, U.S. sec. http://www.nytimes.com/2010/02/19/us/19sfvalley.html?_r=2&scp=1&sq=mountain%20view%20the%20new%20company%20town&st=cse.

Vaughan, Diane. 1996. *The Challenger Launch Decision*. Chicago: University of Chicago Press.

Vázquez, M., E. Pallé, and P. Montañés-Rodríguez. 2010. *The Earth as a Distant Planet: A Rosetta Stone for the Search of Earth-Like Worlds*. New York: Springer.

Verma, Vandi. 2006. "Journalist Report." Mars Desert Research Station Crew Reports. April 24, http://web.archive.org/web/20060927083012/http://www.marssociety.org/MDRS/fs05/0424/journ.asp.

Vertesi, Janet. 2015. *Seeing Like a Rover: How Robots, Teams, and Images Craft Knowledge of Mars*. Chicago: University of Chicago Press.

Warwick, Andrew. 2003. *Masters of Theory: Cambridge and the Rise of Mathematical Physics*. Chicago: University of Chicago Press.

Wilford, John Noble. 2000. *The Mapmakers*. New York: Random House.

Wolfe, Audra. 2002. "Germs in Space: Joshua Lederberg, Exobiology, and the Public Imagination, 1958–1964." *Isis* 93 (2): 183–205.

Wood, Denis, and John Fels. 1992. *The Power of Maps*. New York: Guilford Press.

Wynn, Louise. 2004a. "Commander's Log." Mars Desert Research Station Crew Reports. December 21, http://web.archive.org/web/20051226062056/http://www.marssociety.org/MDRS/fs04/1221/cmdr.asp.

Wynn, Louise. 2004b. "Journalist Report." Mars Desert Research Station Crew Reports. February 15, http://web.archive.org/web/20051225170212/http://www.marssociety.org/MDRS/fs03/0215/journ.asp.

Young, M. Jane. 1987. "'Pity the Indians of Outer Space': Native American Views of the Space Program." *Western Folklore* 46:269–79.

Zabusky, Stacia E. 1995. *Launching Europe: An Ethnography of European Cooperation in Space Science*. Princeton, NJ: Princeton University Press.

Zhan, Mei. 2009. *Other-Worldly: Making Chinese Medicine through Transnational Frames*. Durham, NC: Duke University Press.

Zubrin, Maggie. 2006. "Journalist Report." Mars Desert Research Station Crew Reports. January 4, http://web.archive.org/web/20060927081759/http://www.marssociety.org/MDRS/fs05/0104/journ.asp.

Zubrin, Robert. 1994. "The Significance of the Martian Frontier." Originally published in *Ad Astra*, September/October. Republished on National Space Society website. http://www.nss.org/settlement/mars/zubrin-frontier.html.

Zubrin, Robert. 1996. "The Significance of the Martian Frontier." In *Strategies for Mars: A Guide to Human Exploration* (A 96–27659 06–12), Science and Technology Series, vol. 86, 13–24. San Diego, CA: Univelt.

Zubrin, Robert. 2000. "The Significance of the Martian Frontier." In *The Case for Mars VI: Making Mars an Affordable Destination*, edited by Kelly McMillen, 27–37. Boulder: University of Colorado Press.

Zubrin, Robert. 2002. "Dispatch from Mars Base Utah." Mars Desert Research Station Crew Reports. February 7, http://web.archive.org/web/20081202195151/http://www.marssociety.org/MDRS/fs01/0207/cmdr.asp.

Zubrin, Robert. 2004. *Mars on Earth: The Adventures of Space Pioneers in the High Arctic.* New York: Penguin.

Zukin, Sharon. 1993. *Landscapes of Power: From Detroit to Disney World.* Berkeley: University of California Press.

atmospheres. *See under* Earth; exoplanets

atomism, 192

AURA. *See* Association of Universities for Research in Astronomy

Bachelard, Gaston, 186

Basso, Keith, 31, 56, 103, 145

Batson, Raymond, 108–9

Battaglia, Debbora, 16–17

Beattie, Donald, 46

"being there" concept, 57–58, 89, 96, 172

Bell Labs, 124

Blanco, Victor, 164–65

Blouke, Morley, 124

blueberries of Mars, 51–55

Borucki, Bill, 176–77

Briggs, Geoffrey, 93

Bureau of Land Management, U. S., 36

Butler, Paul, 160

Caillois, Roger, 182

Calvino, Italo, 194

Cameron, James, 29

Campbell, Bruce, 160

Carmel Formation, 39, 40

cartography. *See* mapping

Casey, Edward, 13–14, 151–52

CCDs. *See* charge-coupled devices

Centre National d'Etudes Spatiales, 123, 124

Cerro Tololo Inter-American Observatory (CTIO), 149–50, 153, 155, 187, 208n14; control room, 170–72; facilities, 165–70, 167, 169; history of, 162–65; La Serena, 159–62; observing at, 158–59, 172–75

Certeau, Michel de, 104, 106–7

Charbonneau, David, 137, 139, 140

charge-coupled devices (CCDs), 124–25, 127, 129, 160, 170, 174

Code T, 76–77

Coleman, Gabriella, 82, 201n10

concretions, 51, 52–56, 54, 58–59, 61

Convection, Rotation, and Planetary Transits (CoRoT) space telescope, 123–29, 126, 128, 144, 147

Conversations on the Plurality of Worlds (Fontenelle), 193–94

Copeland, Bill, 93–94

Corner, James, 86–87

CoRoT. *See* Convection, Rotation, and Planetary Transits (CoRoT) space telescope

Cosmicomics (Calvino), 194

Cosmos (television series), 88

Crary, Jonathan, 92, 96

Csicsery-Ronay, Istvan, 59

CTIO. *See* Cerro Tololo Inter-American Observatory

cultural geography, 13, 31, 100, 144, 186–87, 199nn7–8

Curiosity (Mars Exploration Rover), 71, 94, 108, 189

Daston, Lorraine, 8, 9

Davis, Morgan, 44–45

"The Distance to the Moon" (Calvino), 194

Dole, Stephen, 155

Doppler. *See* radial velocity

Dymaxion projection, 87

Earth: atmosphere, 11–12; earthshine, 181; as exoplanet, 180–82; mimetic, 182; nostalgia for, 186; ocean-bearing, map, 183; as Pale Blue Dot, 180; as tidally locked with Moon, 143–44

Earth as a Distant Planet, The (Vázquez/Pallé/Montañés-Rodríguez), 181, 183, 185

Eddington, Arthur, 5–6

Edgerton, Samuel, Jr., 129, 146

Edmondson, Frank, 163

Epicurus, 116, 192

ethnography: "being there" in, 172; of computer world-making, 141; extremeophiles and, 17; landscapes and, 31; of mountaintop observatories, 174; of open source, 82; outer space and, 16, 17–18, 163; and place, 16, 140–41; of planetary science, 3, 11, 14–15, 145, 152, 190; transduction in, 208n16

European Southern Observatory, 156

European Space Agency, 116

exobiology. *See* astrobiology

exoplanetary imagination, 110, 114, 118, 150, 181. *See also* planetary imagination

exoplanet astronomy, 1–2, 20–21; analogies and metaphors, 122, 144; Archimedean point and, 184–85; astrobiology and, 191; epistemology of, 140; as expanding field, 3, 115–22, 189–90; gender representation in, 206n1; infinite worlds and, 192; isolating artifacts in, 126–27; naming and language of, 122, 140–44; and relation between image and text, 146–47; unseen objects in, 113, 147; use of space-based telescopes, 150. *See also* Kepler spacecraft

exoplanets: atmospheres, 131–35, 137; data, importance of, 147, 152; defined, 8, 9; Earth-like, search for, 1–2, 136–37, 175, 184; habitability of, 1–2, 150–51, 154; interiors, 135–40; naming conventions, 204n7; representations, 134; seeing, 111–15, 122, 123–40, 147. *See also* 51 Pegasi, GJ 1214b exoplanet, Gliese 581g exoplanet, HD 189733b exoplanet, Kepler-186f exoplanet

exploration, 77, 80, 83, 107–8; use of term, 18

Exploration Technology Directorate. *See* Code T

extrasolar planets. *See* exoplanets

extremophiles, 17

Fabian, Johannes, 14

Fay, Jonathan, 87

Ferguson, James, 14

Feynman, Richard, 134

fieldwork, 33, 39, 40, 45–46, 56–57, 189, 198n8, 208n16; analog, 20, 27, 33, 46, 50, 55–58

51 Pegasi, 115, 177; 51 Pegasi b exoplanet, 115–16, 117, 154

First Light (Rawlings painting), 55, 65

Fischer, Debra, 21, 150, 152, 159; Project Longshot, 155–56, 158, 160, 207n7

Flashline Mars Arctic Research Station, 28–29

Fong, Terry, 76–77

Fontenelle, Bernard le Bovier de, 193–94

Fox, William, 65

Frankfurt School, 96

frontier narrative. *See under* narrative

Gaia hypothesis, 182

Garvin, Jim, 102, 104–5

Geertz, Clifford, 172

geography and habitable zones, 154

geological narrative of Mars, 20, 33, 34–41; astrogeological, 33–34, 41–50; fieldwork and, 56–57

Gilroy, Paul, 10

GitHub, 82

GJ 1214b exoplanet, 135–42

Gliese 581g exoplanet, 130

globe, world as, 114

Goldilocks zone. *See* habitability

Goodwin, Charles, 119

Google Earth, 34, 73, 78, 83–84, 86–88, 98, 102

Google Mars, 96, 98, 99, 102–3, 104, 202n12

Greeley, Ronald, 108–9

mimesis, 182

MOC. *See* Mars Observer Camera

Moon, 6, 194; astrogeological narrative, 41–50; human exploration of, 18, 76; modeling of, 95, 99, 202n12; as tidally locked with Earth, 143–44

Moran, Patrick, 82–83

Morrison Formation, 38–39, 51

Moylan, Tom, 66–67

Multivator (scientific instrument), 92

Musk Observatory, 62

Myers, Fred, 37

narrative, 26, 31–34, 37; areological, 33–34, 50–59; astrogeological; 33–34, 41–50; frontier, 46–49; geological, 20, 33, 34–41; utopian, 17, 34, 66–70, 200nn20–21; science fiction, 34, 59–66, 200n21

NASA, 29, 116, 201n5; Ames Research Center, 20, 75–76, 82; Constellation program, 76; Discovery program, 176–77; "Exoplanet Community Report," 155; frontier narrative and, 46–49; Goddard Space Flight Center, 72, 142; Jet Propulsion Laboratory, 90; Star and Exoplanet Database, 125–26

National Geographic (magazine), 175

National Science Foundation, 154, 159

National Space Society, 49

Nature (magazine), 53, 132

Navajo Sandstone formation, 53

NewSpace, 18

New York Times (newspaper), 135

nostalgia, 186, 194

Nye, David, 46–47

observatory, role of, 21, 152–53, 158. *See also* Archimedean point; specific observatories

O'Connor, Joe, 48

Olson, Valerie, 16–17

open source technology, 81–83

Opportunity (Mars Exploration Rover), 53, 54–55, 94

otherness, 182–83

outer space: humanness and, 184, 194; placing, 22–23, 196; politics of, 201n3; sociality of, 16–19; terrestrial locality of, 29, 31, 163

parallel universes, 192

Pathfinder (Mars lander/rover), 6, 81, 109, 201n8

pedagogy. *See* visualizations: learning to see

Phoenix (Mars lander), 73, 86, 105

photometry, 123–24, 160, 176

Pinochet, Augusto, 164

Pintupi place-making, 37–38

place-making, 2–3; cosmic loneliness and, 195; difference-making and, 17; habitation and, 19–20, 150; language and, 140–44; mappings and, 74, 84–85; narration in, 19, 37; ontologies of, 13; planetary imagination and, 190–91; on planetary scale, 12; practices, 30; science production and, 15–16; time and, 31; visualizations and, 30, 118–19, 140

place(s): in anthropology, 5, 13, 14; in cultural geography, 13, 31; cultural practices and, 14–15; desire for, 185–86; double exposure of, 30–31, 32, 56; in ethnography, 16; importance of, 190; knowing and, 2–3, 14; in planetary imagination, 145; in planetary science, 9, 11, 22–23, 190; problems and promises of, 13–16; production of science and, 15–16, 17; in science and technology studies, 13, 15–16; time and, 31–32

place-world, use of term, 31, 33, 56, 59, 103, 113–14

Printed and bound by CPI Group (UK) Ltd, Croydon, CR0 4YY

27/10/2024

14580226-0002